NMR Basic Principles and Progress
Grundlagen und Fortschritte 16

Editors: P. Diehl E. Fluck R. Kosfeld

Paul S. Pregosin Roland W. Kunz

^{31}P and ^{13}C NMR of Transition Metal Phosphine Complexes

With 26 Figures

Springer-Verlag
Berlin Heidelberg New York 1979

Dr. Paul S. Pregosin
Dr. Roland W. Kunz

Laboratorium für anorganische
Chemie der ETH
Universitätsstraße 6/8
CH-8006 Zürich

ISBN-13: 978-3-642-48832-0 e-ISBN-13: 978-3-642-48830-6
DOI: 10.1007/978-3-642-48830-6

Library of Congress Cataloging in Publication Data. Pregosin, P. S. ^{31}P and ^{13}C NMR of transition metal phosphine complexes. (NMR, basic principles and progress; 16) Bibliography: p. Includes index. 1. Nuclear magnetic resonance spectroscopy. 2. Transition metal compounds. 3. Phosphine. 4. Complex compounds. I. Kunz, Roland Werner, 1949- joint author. II. Title. III. Series. QC 490.N2 vol. 16 [QD 96.N8] 538'.3s [546.6] 78-31886

Softcover reprint of the hardcover 1st edition 1979

2152/3140-543210

Foreword

For many years coordination chemists have used phosphorus ligands to help stabilize a variety of oxidation states of transition metals. Despite the increasing interest in the chemistry of these complexes relatively few groups have routinely employed ^{31}P NMR as an analytical tool. Additionally, the number of reports concerned with the ^{13}C NMR parameters of these molecular types represents only a small fraction of the phosphorus studies. The growth and commercial development of the so-called Fourier transform NMR spectrometer has sufficiently altered this picture to warrant a presentation concerned with what the coordination chemist can hope to achieve using these NMR methods. Consequently, we have attempted to outline what the practicing chemist would need to know to efficiently make use of these methods and at the same time provide an overview of how these parameters can be used to assist in the solution of current chemical problems.

Zürich, July 1978 Paul S. Pregosin and Roland W. Kunz

Table of Contents

List of Editors

Structural Abbreviations

p-Tol = Tol	= $p\text{-}CH_3C_6H_4$
o-Tol	= $o\text{-}CH_3C_6H_4$
m-Tol	= $m\text{-}CH_3C_6H_4$
Me	= methyl
Et	= ethyl
Pr^n	= n-propyl
Bu^n	= n-butyl
Bu^t	= t-butyl
Bz	= benzyl
Cy	= cyclohexyl
dppm	= $(Ph_2P)_2CH_2$
dppe	= $(Ph_2PCH_2)_2$
dppp	= $(Ph_2PCH_2)_2CH_2$
tppp	= $(Ph_2PCH_2CH_2CH_2)_2PPh$
triphos	= $CH_3C(CH_2PPh_2)_3$
QP	= $P(o\text{-}PPh_2C_6H_4)_3$
TP	= $PPh(o\text{-}PPh_2C_6H_4)_2$

A Motivation and Methodology

I Introduction

For almost a quarter of a century the words "nuclear magnetic resonance" were synonymous with proton measurements. During this period the literature abounded with a seemingly infinite variety of ^{1}H NMR studies concerned primarily with carbon chemistry. Occasionally a "novel" nucleus was studied and, even in those early days, the potential offered by ^{13}C, ^{14}N, ^{31}P and ^{19}F was clearly recognized. Despite the allure, the technical difficulties involved in measuring some of these nuclei were far from trivial. Small magnetic moments and low natural abundance in combination with spin-spin coupling from other nuclei, mostly protons, resulted in a signal-to-noise problem whose severity effectively excluded the study of metal complexes with unfavorable solubility characteristics. The first important breakthrough came with the advent of broad band ^{1}H-decoupling. For example, the featureless broad ^{31}P resonance associated with the commonly used ligand triphenyl phosphine is converted to a sharp, more readily observed singlet when wide-band decoupling is employed (see Fig. 1). Despite this improvement investigation of more interesting molecules, such as catalytically active complexes was forced to await the development of Fourier Transform methods since only with relatively rapid signal averaging methods could sufficient signal-to-noise ratios be achieved.

Considering the expense and technology required before one can routinely measure nuclei such as ^{31}P and ^{13}C it is reasonable to ask what advantages relative to ^{1}H NMR these studies offer to the chemist and in particular to the chemist studying the interactions of phosphine ligands with transition metals. One obvious asset, demonstrated in Fig. 2, is concerned with spectral simplicity. Although the ^{1}H NMR spectrum of *trans*-$PdCl_2(PBu_3^n)_2$ contains, potentially, a wealth of information, the difference in ^{1}H chemical shifts in combination with complications produced by proton-proton and phosphorus-proton coupling

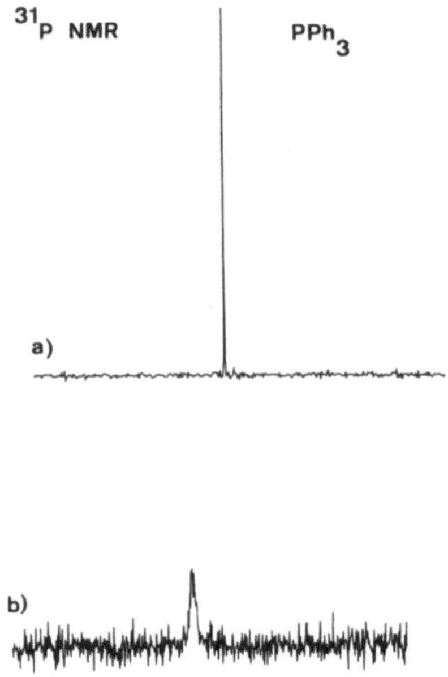

Fig. 1. ^{31}P NMR spectrum of PPh_3 a) ^1H decoupled b) without ^1H decoupling

makes this spectrum essentially useless![1] The ^{13}C NMR spectrum, on the other hand, shows individual groups of resonances for each carbon, generally with well resolved phosphorus-carbon couplings. This is clearly advantageous if one were interested in knowing whether a specific carbon center had somehow changed its chemical environment. Simpler yet, (not shown in the figure), is the ^{31}P NMR spectrum which shows the expected single signal in the presence of broad band ^1H decoupling. This simplicity is particularly advantageous for metal complexes containing more than one type of phosphorus ligand. Thus, although the ^1H and even the ^{13}C NMR spectra for the complex RhCl $(P(OCH_2)_3CCH_2CH_3)(PP)$, I, are not first order the $^{31}P\{^1H\}$ spectrum [*1*] is elegant in its triviality (see Fig. 3). The two equivalent phosphorus atoms appear as a doublet of doublets due to coupling with ^{103}Rh (I = 1/2, natural abundance = 100 %) and the third phosphorus, whereas the single phosphite phosphorus appears as a doublet of triplets stemming from coupling to the metal and the two equivalent phosphorus atoms.

The second asset offered by both ^{13}C and ^{31}P concerns the sensitivity of both the chemical shift, δL, and the coupling constant, $^nJ(M,L)$, (L = ^{13}C or ^{31}P) to structural change. Whereas the proton

---

1. It is reasonable to assume that there exists some H_0 field, generated by a super conducting magnet, which will invalidate this statement.

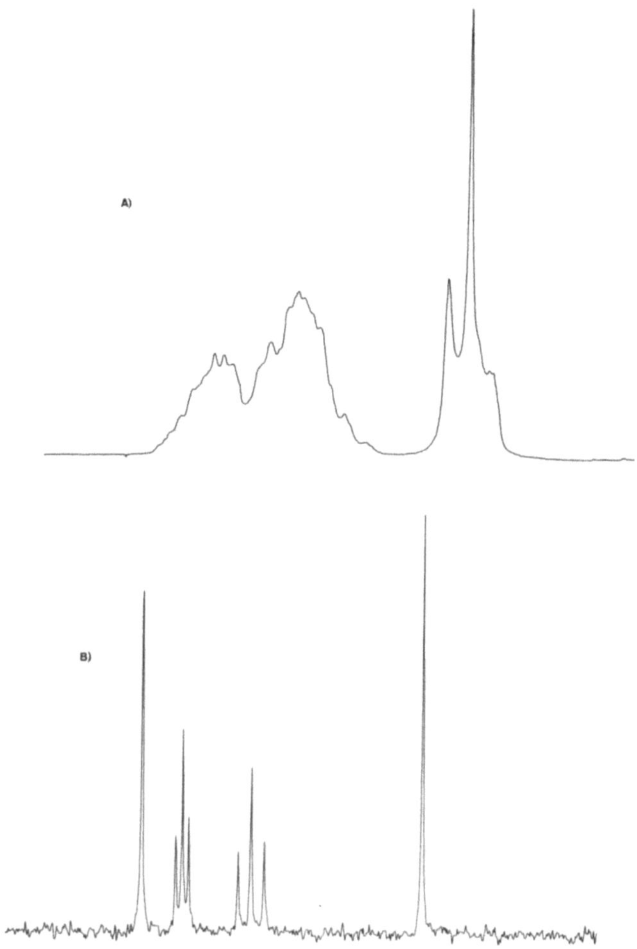

Fig. 2. The ^{1}H(top) and ^{13}C (bottom) NMR spectra stemming from *trans*-$PdCl_2(PBu_3^n)_2$. The $^{31}P\{^{1}H\}$ spectrum is a single line at $\sim \delta$ 10.0 (H_3PO_4) ppm

chemical shift range (excluding metal-hydrides) is often of the order of 10 - 15 ppm, the corresponding values for both ^{13}C and ^{31}P are of the order of 200 ppm. Effectively this means that molecules of similar, but different constitution are more readily distinguished using the heavier nuclei. This is especially useful since nuclei which are situated closer to the reaction site, often show larger changes in their NMR parameters when the molecule undergoes some chemical transformation. We do not mean to suggest that ^{1}H NMR methods have lost their classical importance for this area of chemistry, but rather that equally powerful methods are available whose potential is only just being realized. It is with this point of view that we have

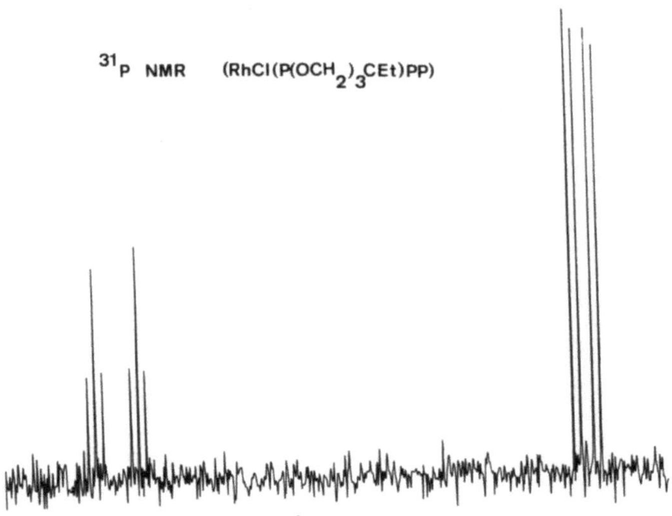

Fig. 3. The $^{31}P\{^1H\}$ spectrum of $[RhCl(P(OCH_2)_3CEt)(PP)]$ showing phosphorus-metal and phosphorus-phosphorus coupling

attemted to review primarily the ^{31}P and to a lesser extent the ^{13}C NMR literature involved with transition metal complexes containing phosphine ligands. We have attempted to cover the literature for the period 1970 - 1977; however, we have no doubt that some reports have escaped our notice. Throughout we have drawn extensively from studies performed in our laboratory and beg the reader's indulgence should he find this emphasis too obvious.

Before proceeding further a few words concerned with the ^{31}P measurements themselves seem à propos.[2]

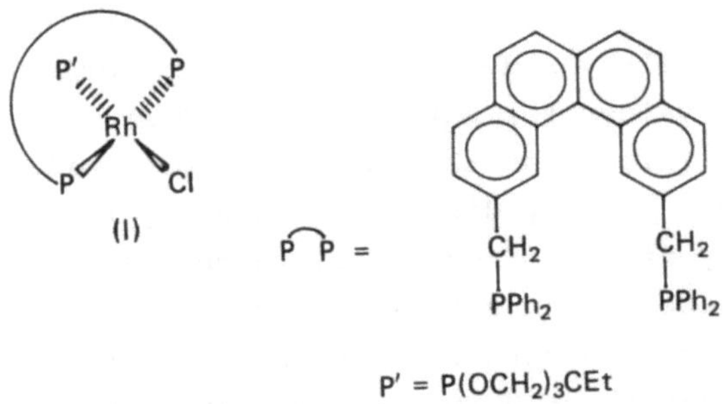

2. The methodology involved in ^{13}C NMR has been extensively reviewed. See for example Levy G., (ed.): Topics in Carbon-13 NMR Spectroscopy, Vols. 1 and 2, New York: Wiley Interscience.

II Measurement Techniques

1 Double Resonance

Early ^{31}P NMR studies were often performed on non-rotating samples containing neat liquids in 8 - 12 mm O.D. tubes. The larger size sample tube was critical since the sensitivity of ^{31}P relative to ^{1}H is less by a factor of approximately 16. Despite this problem Van Wazer and co-workers were able to compile an enormous amount of ^{31}P data [2]. Unfortunately, there is relatively little in this compilation concerned with metal complexes of phosphorus ligands since these are often only sparingly soluble. One early solution to the problem of signal-to-noise involved the so-called INDOR (Inter Nuclear Double Resonance) methodology [3]. In this approach one takes advantage of the relative sensitivity of the ^{1}H (or ^{19}F) nucleus and monitors a sharp line of frequency f, often a methyl proton frequency to which the phosphorus is coupled, while simultaneously sweeping through the phosphorus spectrum with a decoupling frequency, f_2. If the monitoring frequency is sitting "atop" one of the methyl proton resonances, the intensity of this line will change when the correct transitions in the ^{31}P spectrum are selectively irridiated with a suitable decoupling field strength. Each time the decoupling frequency reaches an appropriate ^{31}P transition there will be a response in the ^{1}H spectrum, thus providing an accurate reflection of the ^{31}P NMR spectrum. In addition to allowing the determination of the ^{31}P chemical shift this double resonance method provides a means by which the relative signs of coupling involving ^{31}P can be determined. The primary advantage of the INDOR technique lies in its use of relatively sensitive nuclei to obtain data on less receptive systems.

The principle shortcoming of this method is that it requires that the molecule contains a readily observable resonance to which the phosphorus atom is coupled. Unfortunately, this is not always convenient.

2 Fourier Transform NMR

The earliest efforts at combating the signal-to-noise problem involved signal accummulation in a small computer or CATing (Computer of Average Transients). The spectrum was measured, the result stored, and the process additively repeated until such time as sufficient signal-to-noise could be obtained. This procedure is efficient, provided that the time involved for a single scan is not too long, since there is often a practical limit upon the amount of time that will be allowed for a single spectrum. A more efficient way of accummulating signals employs pulsed NMR techniques. This methodology takes advantage of the long recognized fact that the *frequency response* of an ensemble of nuclear spins to a short relatively high power radio frequency pulse is related to the normally observed *frequency spectrum* via a simple Fourier Transformation [*4*]. Since the pulse can be created such that it excites a relatively wide range of frequencies (generally of the order of $1/\tau_P$ Hz, τ_P = pulse length in seconds) we can irradiate the entire ^{31}P range of chemical shifts essentially instantaneously. At time zero the ^{31}P equilibrium magnetization vector M, (the sum of all the individual nuclear magnets) will appear as in Fig. 4a.

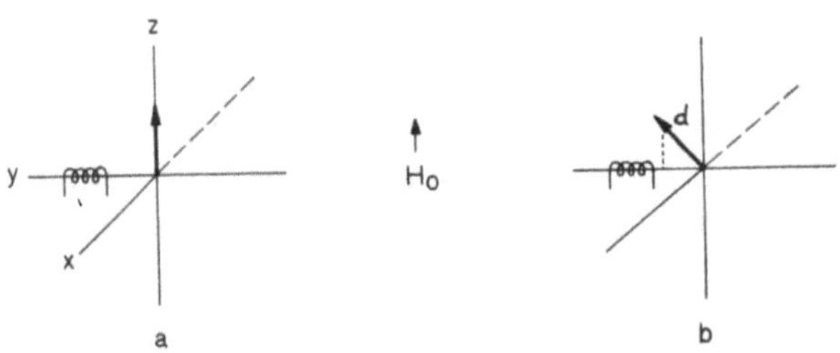

Fig. 4. The magnetization vector a) at equilibrium. b) after a radiofrequency pulse. There is now a component of magnetization in the XY plane which can be detected by the receiver coil

After the pulse the magnetization will be tipped away from the equilibrium direction by some angle, α, the flip angle, such that $\alpha = \gamma H_1 \tau_P$ where H_1 represents the strength of the transmitter field, τ_P the length of time the pulsed field is "on" and γ is the gyromagnetic ratio of the nucleus in question, in this case ^{31}P. Since we have created magnetization in the XY plane a signal will be induced in the receiver coil as in Fig. 4b. Immediately following the pulse the spin-lattice, T_1, and spin-spin, T_2, relaxation processes will begin to restore the system to its previous equilibrium position. Since the magnetization in the direction of the receiver coil is decaying in the absence of any perturbation (the pulse has been shut off) the pattern detected by our receiver is called a Free Induction Decay or

FID. The FID which results when a single type of ^{31}P nucleus is excited (e.g. a proton decoupled ^{31}P singlet) is shown in Fig. 5a for $P(OMe)_3$ and appears as an exponentially decaying sine wave whose frequency is the difference between the center frequency of the excitation pulse (sometimes called the carrier wave) and the Lamor frequency of our signal. For the case where several absorptions occur the decay will appear somewhat differently such as in Fig. 5b for which the ^1H

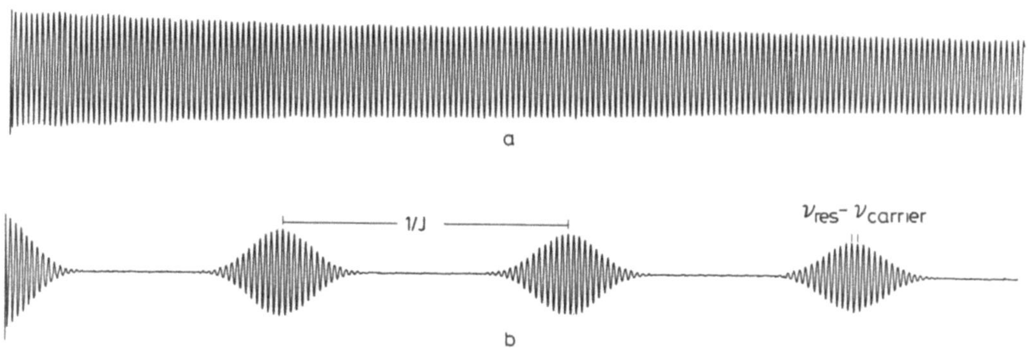

Fig. 5. The ^{31}P NMR spectrum of $P(OMe_3)$ a) ^1H decoupled and b) without decoupling

decoupler has been switched-off. This is still a relatively simple FID from which both ν and $^3J(P,H)$ may be directly calculated. Often the FID appears as the superposition of a number of simultaneously decaying frequencies such as in Fig. 6. Fourier transformation of this time domain data produces a frequency domain spectrum which, after phase correction, can be plotted in the conventional manner. Although the frequency domain data is often the goal, the time domain data should not be ignored since they often can provide useful information. The FID is a measure of the spin-spin relaxation, T_2^* [3], and as such is a direct read-out of the line widths in the spectrum. Rapidly decaying FID's signify relatively broad resonances whereas longer lasting decays indicate the presence of sharper lines. A knowledge of the line widths to be expected can be of further use in that it is possible to mathematically operate on the FID so as to either increase the signal-to-noise or the resolution in the transformed spectrum. The reader is recommended to consult the literature [5] for the details of these operations. Given that all of the necessary spectral infor-

3. T_2^* is related to the observed line widths and contains, in addition to the spin-spin relaxation times, T_2, a contribution from the magnetic field homogeneity. Thus $1/T_2^* = 1/T_2 + (\gamma H_0/2)$

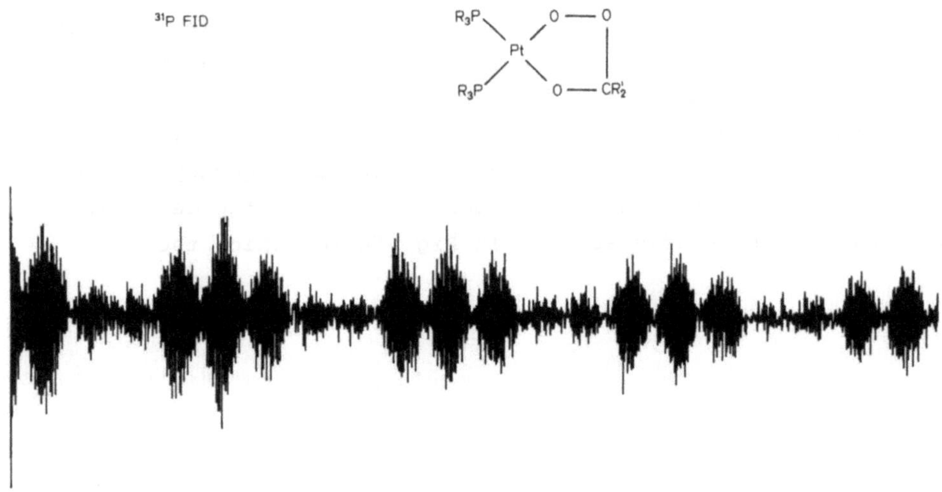

Fig. 6. The $^{31}P\{^1H\}$ FID stemming from *cis*-Pt($(CH_3)_2CO_3$)$(PCy_3)_2$. D. Grove and L. M. Venanzi 1978

mation is contained in the FID it is clear that data accummulation in this mode is especially efficient since the time required to collect the "spectrum" is often of the order of T_2* or less. Thus we have collected in seconds (or as for ^{13}C and ^{31}P, fractions of a second) the same information which might normally have required minutes to acquire in the continuous wave mode!

It is well known that repetitive scanning through an NMR spectrum without allowing sufficient time between scans will produce saturation phenomena. Conceptually the same problem exists in the pulse mode since we are summing many free induction decays. Normally, we pulse, acquire data for about a second[4], pulse again ... etc. Subjecting the system to a train of pulses each of which tip the magnetization vector 90° away from the Z-axis (so-called 90° pulses) will only be effective if the acquisition time is at least three times as long as T_1, the spin lattice relaxation time (4 or 5 T_1's would be better) since we then will have restored the magnetization vector to its equilibrium position. Since T_1 values for both ^{31}P and ^{13}C vary considerably, but are often of the order of seconds and tens of seconds, the acquisition time is usually not long enough to permit the

4. The reader should consult Refs. 4 - 6 for an introduction to the applications of sampling theory within this context. Suffice it is to say that acquiring 5000 Hz of information in 8192 data points requires approximately 0.8 seconds.

use of 90° pulses. Normally one chooses to reduce α, the flip angle, as opposed to waiting, since the time of the measurement is important. Calculations concerned with the selection of an optimum value for α have been made by Ernst and co-workers [6].

The primary advantages stemming from Fourier Transform NMR are concerned with time. Data can be accummulated relatively quickly while covering spectral widths of the order of 5,000 - 25,000 Hz within a fraction of a second; however, to do this effectively some thought must be given to relaxation phenomena and in particular, T_1. There are almost no T_1 data for phosphine complexes and very little for phosphorus containing molecules in general [7]. Since this parameter has a practical significance for all wishing to measure ^{31}P and ^{13}C by F.T. methods (and this means almost everyone doing work in phosphorus chemistry) it is worth closer attention.

III Spin-Lattice Relaxation Times

In the previous section we mentioned that the magnetization, M, can be tipped away from the Z-direction by an angle $\alpha = \gamma H_1 \tau_p$, where H_1 is the strength of the transmitter field; however, no mention was made of the processes involved in restoring the magnetization to its equilibrium position in the Z-direction. The time for these processes, which involve the interactions of M with the random magnetic fields generated by the motions of the atoms and electrons within the molecules, is called the spin-lattice relaxation time, T_1.

There are several ways in which this parameter can be measured; however, the most commonly employed technique involves some form of the inversion recovery methodology [8]. This involves the application of a 180° pulse (see Fig. 7) which by definition, rotates the ^{31}P magnetization vector from the positive to the negative Z-direction. After a short waiting time, τ, during which relaxation occurs, a 90° pulse is applied. The magnetization vector will now possess a component in the direction of the receiver coil. If τ was long enough such that M had a component in the + Z-direction, the signal appears positive. For relatively short τ values negative signals are observed. After a sufficient time, T, is allowed for the spin system to completely reequilibrate the entire process can be repeated "n" times such that sufficient signal-to-noise can be obtained. A typical $(180-\tau-90-T)_n$ sequence for the complex is shown in Fig. 8. The problems involved in this and related measurements have been discussed by several authors [8]. Since the amplitude of the measured signal, A, can be expressed as shown in equation (1), T_1 can be obtained by plotting the difference between $\ln(A_0)$ and $\ln(A)$

$$\ln(A_0-A) = \ln 2A_0-\tau/T_1 \qquad (1)$$

against τ (where A_0 is the intensity corresponding to a 90° pulse).

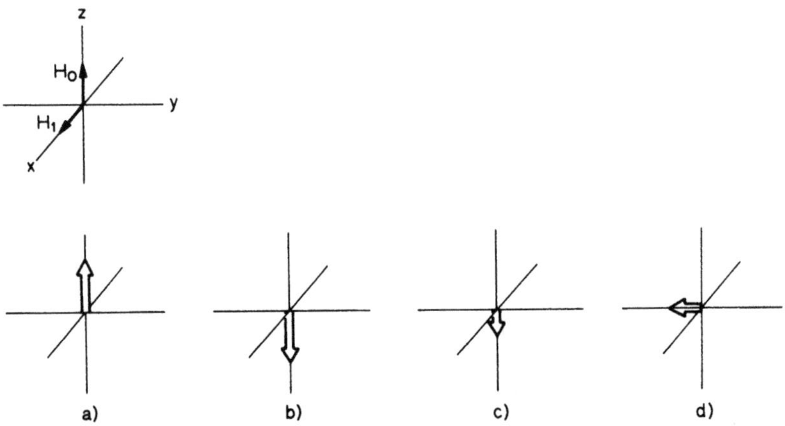

Fig. 7. The magnetization vector at various times during a T_1 (inversion-recovery) experiment a) at equilibrium b) after the 180° pulse c) after the waiting time, τ d) after the 90° pulse

The relaxation mechanisms thought to be operating for ^{31}P are:
a) dipole-dipole b) spin-rotation and
c) chemical shift anisotropy [7].

For organophosphine derivatives the dipole inducing ^{31}P relaxation will stem from the proton(s) on the organic portion of the molecule. The relation expressing the dipolar contribution to the ^{31}P spin-lattice relaxation is shown in Eq.(2), where

$$1/T_{1_{d,d}} \propto \sum_i \gamma^2_H \gamma^2_P \tau_c / r_i^6 \qquad (2)$$

τ_c is the correlation time for the reorientation of the $^{31}P-{}^1H$ vector, r is the distance separating the dipoles and the γ's are the nuclear gyromagnetic ratios.

For the spin-rotation mechanism, where the magnetic fields which relax the phosphorus are generated from the motion of the molecular magnetic moment, the appropriate equation is

$$1/T_{SR} \propto ITC_{eff}^{2}\tau_J \qquad (3)$$

where I is the moment of inertia of the molecule, T is the temperature, C_{eff}^{2} the spin rotation tensor and τ_J the angular momentum correlation time.

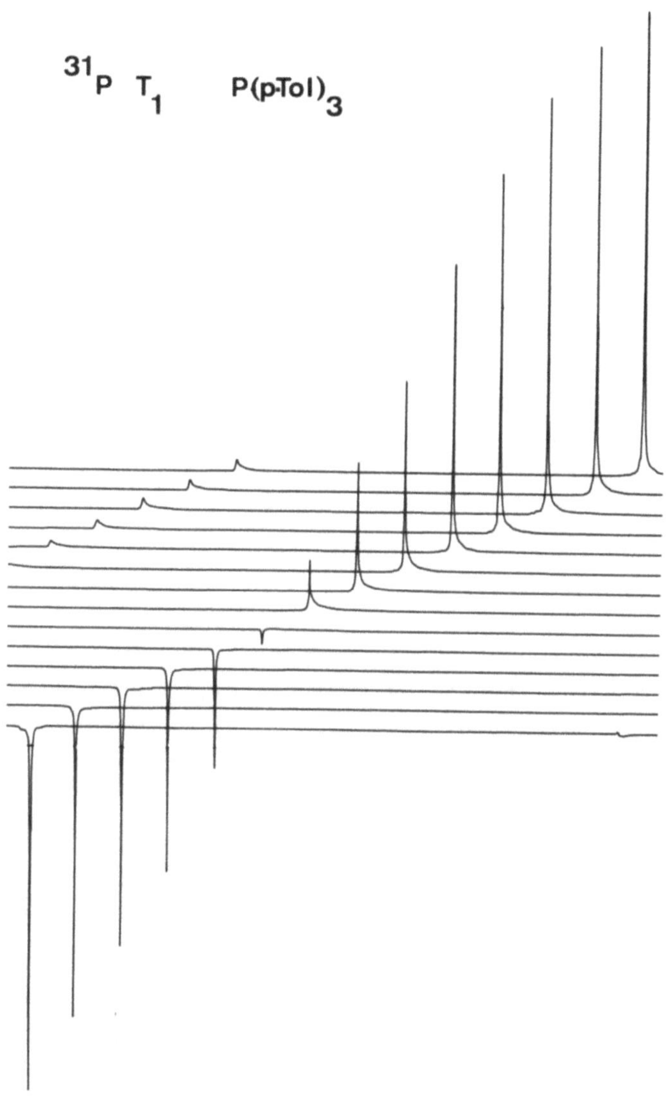

Fig. 8. Inversion-recovery sequence for $P(p\text{-}Tol)_3$. As the waiting time, τ, increases the signal will become increasingly positive

The chemical shift anisotropy relation, shown in Eq.(4)

$$1/T_{csa} \propto H_0^{\,2}(\sigma_{\parallel}-\sigma_{\perp})^2\tau_c \tag{4}$$

contains the magnetic field strength, the screening tensor parallel and perdendicular to an assumed axial symmetry axis and the correlation time, τ. This mechanism is generally not as important as the first two.

In Table 1 (see Chap. G) are shown ^{31}P T_1 values for some phosphines and some phosphine complexes of Pd(II). The most obvious point concerns the change in T_1 as a function of the length of the alkyl chain in the trialkyl phosphine derivatives. In addition, one can see from Table 1 that there is a slight dependence of the ^{31}P T_1 value upon the complex geometry with *cis*-complexes showing slightly smaller spin-lattice relaxation times than the corresponding *trans*-isomers. It is tempting to seek an explanation for these latter data in terms of intra-molecular non-bonding interactions between the two phosphine ligands; however, further studies will be required before this hypothesis can be considered seriously. A particularly important point for practitioners is the observed solvent effect on the ^{31}P T_1. For the complexes $PdCl_2P_2$(P = various monodentate phosphines) one finds T_1 values as much as fifty per cent *longer* in CD_2Cl_2 as in $CDCl_3$. Thus these two often interchanged chlorohydrocarbon solvents can drastically effect the ease with which one obtains a ^{31}P spectrum. A number of other ^{31}P relaxation studies have concerned themselves with phosphates, phosphites, trihalophosphines and phosphonium salts [*7*].

Although all three mechanisms have been suggested as contributing to T_1 relaxation of uncoordinated phosphorus derivatives, complexation of a phosphine to a second or third row transition metal is accompanied by a relative increase in the dipolar contribution. The coordination of the phosphorus to the metal can bring about a significant increase in the correlation time, τ. Since the complex will be less mobile than the free ligand, and as $T_{1_{DD}}$ is inversely proportional to τ_c (in the extreme narrowing condition), an increase in τ_c will decrease $T_{1_{DD}}$. A further hint that the dipole-dipole mechanism is important for coordinated ^{31}P comes from the observed S/N increase when the sample temperature is lowered [*9*]. This is readily understood in terms of a change in the viscosity of the solution and the resulting effect on τ_c. Clearly additional knowledge concerned with the ^{31}P T_1 can lead to a more efficient use of spectrometer time.

IV The ^{31}P Nuclear Overhauser Effect

The application of broad-band ^{1}H decoupling techniques in conjunction with the direct observation of ^{31}P offers several advantages simultaneously. Firstly a potentially complex spectrum is simplified while enhancing signal-to-noise, secondly, since two frequencies (^{31}P and ^{1}H) are being employed, the possibility for the determination of the relative signs of coupling constants is retained and thirdly, a significant increase in the signal intensity via a nuclear Overhauser enhancement, NOE, is introduced [*10*]. Since anything which facilitates obtaining the NMR spectrum is important, points 1 and 3 are especially commendable with the latter warranting a few additional words.

It is well known that the energy levels for a simple two spin system AX (in this case A = ^{1}H, X = ^{31}P) can be represented as shown in the scheme.

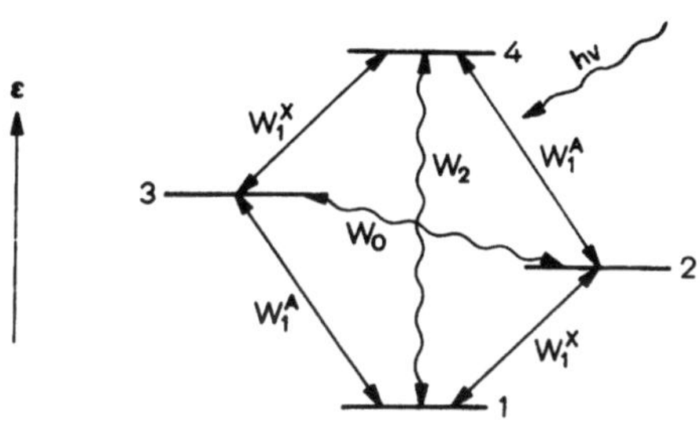

Irradiation of both A transitions, 2 → 4, and 1 → 3 increases the populations of 3 and 4 by exactly the amount lost from levels 1 and 2. There will therefore, be no net gain in intensity for the X-transitions. Since irridiation of a particular A transition (e.g. 2 → 4 as shown by the arrow in the figure) only selectively enhances the 1 → 2 X-transition (population level 2 depleted) at the expense of the 3 → 4 X-transition (population of level 4 increased) it is clear that the pathways W_0 and W_2 must be important if we are to find a signal enhancement. Specifically, the W_2 process, which increases the population of level 1 while decreasing that of level 4, must be more effective than the W_0 process. Should this be the case the intensity of both X-transitions will increase. Solomon [*11*] has shown that the enhancement to be expected can be expressed as shown in Eq.(5).

$$\text{enhancement} = [(W_2-W_0)/(2W^A+W_2+W_0)](\gamma_A/\gamma_X) \qquad (5)$$

For the case where the extreme narrowing condition ($\tau\omega << 1$) is fulfilled and the ^{1}H-^{31}P dipol-dipole interaction is the dominating spin-lattice relaxation mechanism, Eq.(5) can be shown to reduce to the simple expression shown in Eq.(6)

$$\text{enhancement} = [(\gamma_A/\gamma_X)/2] \tag{6}$$

Since the ratio of gyromagnetic constants, $(\gamma_{^1H}/\gamma_{^{31}P})$, is ∿ 2.47, Eq.(6) predicts possible enhancement of the ^{31}P NMR resonance of 124%! (1 + enhancement = 224%). This is clearly a significant increase and if achieved might provide the difference between readily detecting, or just barely observing, the desired signal. The maximum theoretical NOE values for other nuclei such as ^{13}C and ^{15}N are 2.98 [*12*] and - 3.98 [*13*] respectively, and these are routinely observed for molecules where dipole-dipole relaxation dominates. It is possible that other mechanisms will play a role in ^{31}P relaxation in which case the NOE will be smaller than the theoretically expected value by an amount dictated by Eq.(7),

$$\text{NOE} = 1.24(T_{1_{OBS}}/T_{1_{DD}}) \tag{7}$$

where $T_{1_{DD}}$ is the dipolar contribution to the phosphorus spin lattice relaxation time and $T_{1_{OBS}}$ is the observed spin lattice ralaxation time. There is relatively little information concerning nuclear Overhauser effects in ^{31}P NMR in general and essentially nothing for metal complexes. Even so, the entire spectrum of NOE possibilities has already been observed (see Tables 1 and 2 Chap. G).

In Table 1 we show some NOE values for a series of Pd(II) complexes in which we see that significant signal enhancement is observed relative to the "coupled" spectrum. Thus proton decoupling of the ^{31}P spectrum in metal complexes can indeed bring a significant bonus. For the complex *trans*-$PdCl_2(PBu_3^n)_2$ the nuclear Overhauser effect is almost complete (1.20) and therefore the ^{31}P relaxation in this complex is essentially dipolar. It is worth noting that a full NOE does not necessarily require short T_1 values. The conclusion to be drawn from a complete signal enhancement is that other mechanisms, e.g. spin rotation, are not efficient in relaxing the phosphorus atoms; however, we note that in some of the Pd(II) complexes in Table 1 the increasing NOE-values are associated with decreasing ^{31}P T_1 values (e.g. in CD_2Cl_2 T_1 = 14.7, 9.5 and 6.0 for PEt_3 PPr_3^n and PBu_3^n complexes, respectively). The interest in this ^{31}P NMR parameter will undoubtedly grow since there is much physical information to be obtained via such measurements.

B Coupling Constants

I One-Bond Metal-Phosphorus Coupling Constants

The spin-spin scalar coupling between two nuclei has long since become a routine tool in the armament of chemists. This stems partially from the enormous number of empirical relations routinely in use as structure and hybridization probes and partially from the theoretically more tractable relationships which are thought to correctly describe this spin-spin interaction. There are several levels of sophistication commonly in use when discussing the theory of the one-bond metal-phosphorus coupling constant, $^{1}J(M,P)$, and this topic, in combination with the question of metal-phosphorus bonding, will be considered first in this chapter. This will be followed by a short presentation concerned with the question of the two-bond through-metal coupling $^{2}J(P,P)$.

1. Theory

The interactions which lead to a nuclear spin-spin coupling in fluids were first formulated by Ramsey [*16*] in terms of a contact (Fermi) mechanism, a spin-orbital- and a spin-polarization mechanism. Both McConnell [*17*] and Pople and Santry [*18,19*] have described modifications of Ramsey's equations which are suitable for use with molecular orbitals derived from LCAO methods. The Pople and Santry treatment has an advantage in that it is capable of producing both negative and positive coupling constants. The McConnell use of the mean excitation energy approximation, which we shall mention later, necessarily results in the same relative sign for all $^1J(X,Y)$ values, thus failing to rationalize the existence of couplings, e.g. $^1J(P,C)$, which are known [*20*] to be both positive and negative signs. Of the three terms mentioned above the spin-dipolar and spin-orbital contributions are generally accepted [*18,21,22*] to be small relative to the Fermi contact term and approach zero as the π-bond order decreases, thus making the theoretical approach considerably simpler. Employing the Pople and Santry approach, the molecular orbital expression for the contact contribution to the spin-spin coupling between directly bonded nuclei A and B is given by

$$J_{AB} = -(64/9)\beta^2\gamma_A\gamma_B \sum_i^{occ} \sum_j^{unocc} ({}^3\Delta E_{i\to j})^{-1} (\psi_i|\delta(r_A)|\psi_j)(\psi_i|\delta(r_B)|\psi_j) \quad (2.1)$$

where β is the Bohr magneton and γ is the nuclear magnetogyric ratio. The terms in the summation relate to the interaction of the triplet excited states with the singlet ground states (to avoid a discrepancy with the Pauli principle) evaluated at the nuclei A and B. In the LCAO approximation this formula becomes

$$J_{AB} = \frac{64\pi^2}{9h}(g\beta\hbar)^2\gamma_A\gamma_B \sum_i^{occ} \sum_j^{unocc}$$

$$({}^3\Delta E_{i\to j})^{-1} \sum_{p\in A} \sum_{q\in B} c_{ip}^s c_{iq}^s c_{jp}^s c_{jq}^s |\Psi_{s,p}(0)|^2 |\Psi_{s,q}(0)|^2 \quad (2.2)$$

with $g = 2.0023$, $\Delta E_{i\to j} = \varepsilon_i - \varepsilon_j$ where ε is the energy of an occupied (ε_i) or an unoccupied (ε_j) orbital level, c_p^s and c_q^s are the LCAO-Eigenvectors of an s-type atomic orbital, centered on A or B and $|\Psi_{s,p}(0)|^2$ and $|\Psi_{s,q}(0)|^2$ are the s-electron densities at the nucleus of the atomic orbital p or q.

If only one-center integrals involving valence s orbitals on A and B are retained then

$$J_{AB} = -\frac{64\pi^2}{9h}(g\beta\hbar)^2\gamma_A\gamma_B|\Psi_{s,A}(0)|^2|\Psi_{s,B}(0)|^2 \sum_i^{occ} \sum_j^{unocc}$$

$$({}^3\Delta E_{i\to j})^{-1} c_{i,A}c_{j,A}c_{i,B}c_{j,B} \quad (2.3)$$

The singlet-triplet excitation energy ${}^3\Delta E_{i\to j}$ is given by

$$ {}^3\Delta E_{i\to j} = e_j - e_i + (\Psi_j^2|\frac{e^2}{r_{12}}|\Psi_i^2) $$

where e_i and e_j are the Hartree-Fock orbital energies. The integral $(\Psi_j^2|\frac{e^2}{r_{12}}|\Psi_i^2)$ represents the coulombic repulsion between the molecular orbitals Ψ_i and Ψ_j and becomes in the LCAO-approximation

$$ (\Psi_j|\frac{e^2}{r_{1,2}}|\Psi_i^2) = \sum_k \sum_l \sum_m \sum_n C_{ki}C_{li}C_{mj}C_{nj}(\emptyset_k\emptyset_l|\frac{e^2}{r_{12}}|\emptyset_m\emptyset_n) $$

In the extended Hückel (E.H.) method used by Pople and Santry this integral is omitted because of the neglect of the interelectronic repulsion.

Finally, if the mean triplet excitation energy approximation is made, Eq.(2.3) can be further simplified and becomes Eq.(2.4)

$$ J_{AB} = (32\pi/9)\ \gamma_A\gamma_B\beta^2({}^3\Delta E)^{-1}|\Psi_{s,A}(0)|^2|\Psi_{s,B}(0)|^2 \cdot P^2_{sAsB} \qquad (2.4) $$

where $P_{sA\ sB}$ is the s_A-s_B-bond order given by

$$ P_{AB} = 2 \sum_i^{occ} C^s_{iA}C^s_{iB} $$

The same equation was derived by McConnell making the mean triplet excitation energy approximation at an earlier stage.

An even more simplified and often invoked form of Eq.(2.3) is shown in Eq.(2.5)

$$ J_{AB} \propto k\alpha_A^2\alpha_B^2|\Psi_A(0)|^2|\Psi_B(0)|^2/{}^3\Delta E \qquad (2.5) $$

In this modification k can be considered as a constant for the set of compounds in question and the α^2-terms represent the localized hybrid bond s-character in a valence bond description. This relation is often encountered in studies concerned with one-bond metal-phosphorus coupling and we shall return to it later on.

To facilitate the comparison of nuclear spin-spin coupling constants in different molecules, it is convenient to introduce a reduced coupling constant K_{AB} which is independent of the nuclear magnetic moments of A and B and depends only on the electronic environment. This can be defined as

$$K_{AB} = (2\pi/\hbar\gamma_A\gamma_B) \cdot J_{AB} \tag{2.6}$$

Combining the more general Eq. (2.3) with (2.6) the reduced one-bond coupling constant can be given by

$$K_{AB} = (64\pi^2/9)|\Psi_{s,A}(O)|^2|\Psi_{s,B}(O)|^2 \cdot \pi_{AB} \tag{2.7}$$

with

$$\pi_{AB} = +4 \sum_i^{occ} \sum_j^{unocc} (E_i-E_j)^{-1} c^s_{i,A} c^s_{i,B} c^s_{jA} c^s_{jB} \tag{2.8}$$

and should, according to these formulae, be dependent on the s-electron density at the nuclei A and B and on the eigenvalues and eigenvectors of the LCAO-MO's.

Since $^1J(M,P)$ values have been reported for ^{51}V [*23,24*], ^{57}Fe [*25*], ^{59}Co [*26*], $^{65,63}Cu$ [*27*], $^{95,97}Mo$[*28*], ^{103}Rh [*29-32*], $^{107,109}Ag$ [*33*], $^{111,113}Cd$ [*34,35*], $^{117,119}Sn$ [*36-38*], ^{183}W [*39-43*], ^{189}Os [*44*], ^{195}Pt [*45-48*], ^{199}Hg [*49-51*] and ^{207}Pb [*36*], one can get a feeling for the validity of the relationship between $|\Psi(O)|^2$ of the metal and the direct coupling constant by looking at some data.

In Fig. 9 we present a plot of $|\Psi_{S,M}(O)|^2$ in atomic units [*52*] versus the mean values of the one-bond coupling constants $^1K_{MP}$. The lengths of the bars are proportional to the spread of the actual values.[5]

Although the deviation from linearity is significant, Fig. 9 qualitatively reflects the dependence of $^1K(M,P)$ on the s-electron expectation values at the metal. With metals such as Pt and Hg $^1J(M,P)$ values are of the order of KHz, whereas for Rh these values are usually < 200 Hz. For Rh, W and Pt phosphine complexes relative sign determinations have revealed positive reduced couplings for $^1K(M,P)$; however, $^1K(^{117,119}Sn,^{31}P)$ values, whose magnitudes are sometimes

5. Relativistic expectation values from Ref. 256 were used for this plot.

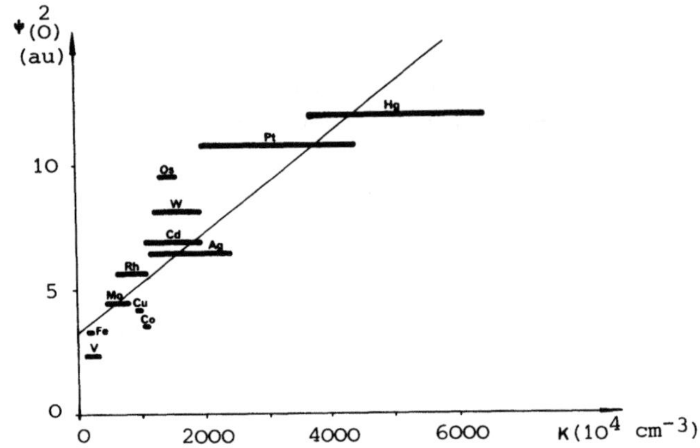

Fig.9. Plot of $\psi(O)^2$ in atomic units against the reduced coupling constant, K

greater than one thousand Hz, are negative [36]. Negative K values have been found for both the free ligand $(Me_3Sn)_3P$ and the tungsten complex $W(CO)_5(P(SnMe_3)_3)$ [53,54].

Although K and not J is the more informative parameter, the chemical literature uses J almost exclusively. Therefore, unless the sign of the coupling has significance J will be given throughout the remainder of the text.

The changes in the values $^1J(M,P)$ for a given metal or, put in terms of Fig. 9, the horizontal lengths of the bars, could result from changes in the polarisability, π_{AB}, or either of the $|\Psi_{ns}(O)|^2$ terms. Although a quantitative analysis of each of these terms is not possible there are some hints available concerning relative contributions. Since the relative magnitudes of s-electron densities at the metal are expected to be directly related to the metal oxidation state (e.g. $Sc^O/Sc^{-1} = 2.4$, $Ti^O/Ti^{-1} = 2.5$, $V^O/V^{-1} = 2.6$, $Cr^O/Cr^{-1} = 2.6$, $Mn^O/Mn^{-1} = 2.5$, $Fe^O/Fe^{-1} = 2.5$, $Co^O/Co^{-1} = 2.8$, $Ni^O/Ni^{-1} = 2.5$, $Cu^O/Cu^{-1} = 2.5$) [55] one might expect to observe a larger value of $^1K(M,P)$ for the more positively charged octahedral Rh(III) and Pt(IV) phosphine complexes relative to the less positively charged square planar Rh(I) and Pt(II) analogs. That this is not the case is shown by structures II-V.

Indeed it is well known that for comparable ligands $^1J(Pt(IV),P) \leq {}^1J(Pt(II),P) \leq {}^1J(Pt(O),P)$. A selection of $^1J(Pt,P)$ values for these three oxidation states are shown in Tables 3-9 (see Chap. G). Despite the fact that these transition metal complexes should not be considered as containing metal ions of "fixed" positive charge, e.g. O, + 2 or + 4 for platinum, the experimental observations suggest that

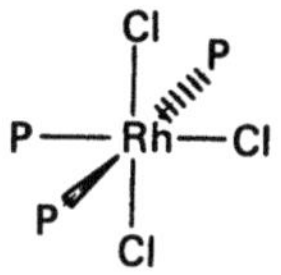

$^{1}J(Rh, P)$ = 114, 86 Hz
P = $PMePh_2$

(II)

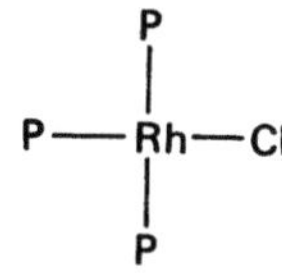

$^{1}J(Rh, P)$ = 189, 142 Hz
P = PPh_3

(III)

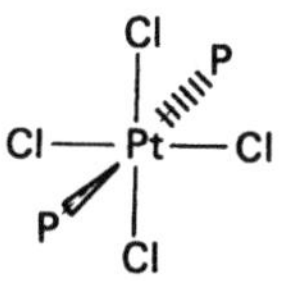

$^{1}J(Pt, P)$ = 1455 Hz
P = PEt_3

(IV)

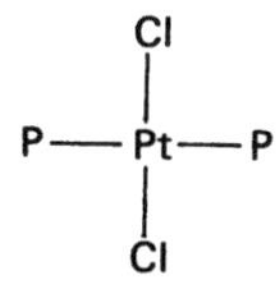

$^{1}J(Pt, P)$ = 2397 Hz
P = PEt_3

(V)

$|\Psi_{ns}(0)|^2_M$ may not be the parameter whose change would explain the various magnitudes of $^{1}J(M,P)$ for a given metal. In addition there have been several reports concerned with the complexes $[PtX(PR_3)_3]^+Y^-$ (see scheme) in which two very different $^{1}J(Pt,P)$ values are found [*45*].

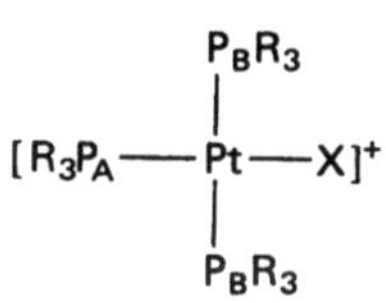

$^{1}J(Pt, P_A)$ = 3 497 Hz
$^{1}J(Pt, P_B)$ = 2 380 Hz
X = ONO_2
R = Et

(VI)

It is suggested that these data support the relative unimportance of the $|\Psi_{Pt}(0)|^2$ term since only one value of this term can be assigned for this Pt complex. If indeed, for a given metal, the $|\Psi(0)|^2$ parameter changes considerably, one can formulate a second type of reduced coupling.

$$F_{AB} = K_{AB}[1/|\Psi_A(0)|^2|\Psi_B(0)|^2]$$

This would also more readily permit comparison of $^{1}J(M,P)$ values for different metals and has been suggested [*56*], although not generally accepted.

A second possibility for the horizontal variations in Fig. 9 is the $|\Psi_{3s}(0)|^2_P$ term. It is well known that for a given type of metal complex changing from a phosphine to a phosphite ligand induces a change in $^{1}J(M,P)$ of between 50 and 100% [*57,58*]. The general case, e.g. substitution of a more electronegative group for carbon on phosphorus, is illustrated nicely by the series of octahedral complexes $W(CO)_5(PX_3)$ [*43*]. In Fig. 10 and in Table 10 (see Chap. G) we show this effect in which the value of $^1J(W,P)$ increases directly as the electronegativity of X. Thus for X = F, Cl, Br and Bu^n, one finds $^1J(W,P)$ values of 485, 426, 398 and 227 Hz, respectively [*43*]. The authors suggest that more than one linear relation is observable from this graph and indeed the $P(halogen)_3$ series stands out. An extension of this type of electronegativity correlation to X groups which are less electronegative than carbon e.g. for X = $GeMe_3$(162.5 Hz) $SiMe_3$ (150.9 Hz) and $SnMe_3$(142.7 Hz) [*59*] has been reported. There is no definitive explanation offered for the observed changes which amount to approximately a factor of 3; however, the $|\Psi_{3s}|^2$ term is given as a possibility. The term $|\Psi(O)|^2$ *was* invoked to explain the same types of changes in some Pt(II) phosphite complexes [*60*]. In this connection it is interesting to note that for the protonated form of the following ligands:

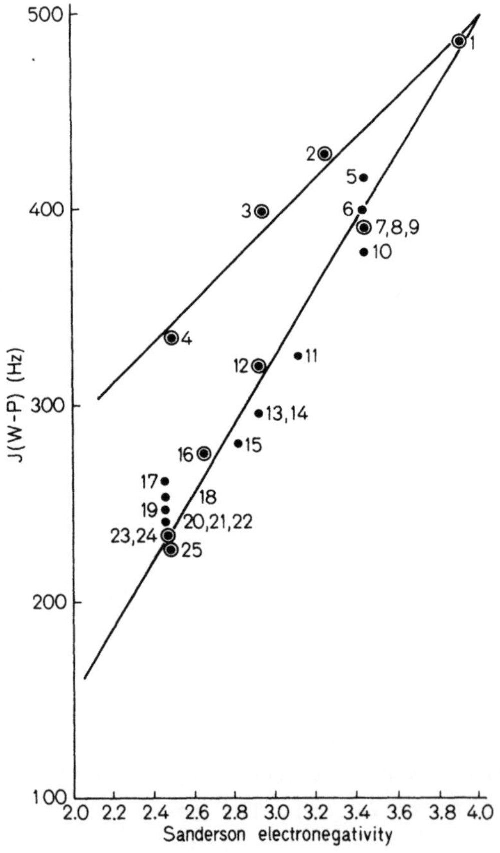

Fig. 10. Plot of $^1J(W,P)$ against the Sanderson electronegativity for the complexes $W(CO)_5PX_3$, with different X groups. Points 1-4 represent the halogens [*43*]

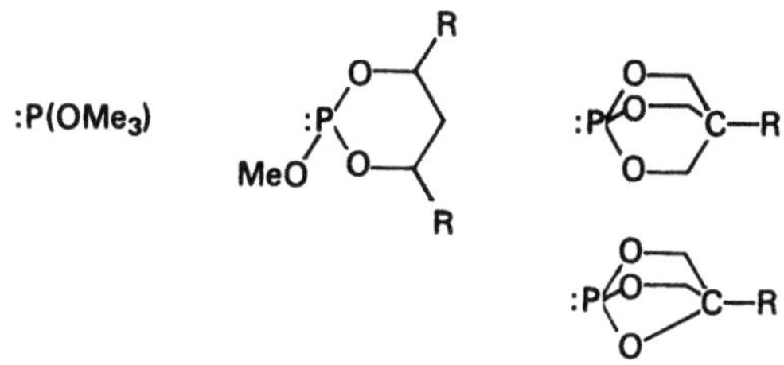

a series of CNDO/2 calculations have shown that $^1J(P,H)$ correlated well with the calculated charge on phosphorus and not with the "s-character" of the phosphite lone pair [*60*].

Although the substitution of an alkoxy or halogen group for an alkyl group invariably increases $^1J(M,P)$ this is not always the case when the more electronegative aryl substituent replaces an alkyl group. Thus, although $^1J(M,P)$ for the complexes $PtCl_2(PR_nPh_{(3-n)})_2$ [*40*] increases with decreasing n, the reverse seems to be true for the metals Sn(IV), Cd(II) and Hg(II) (see Tables 11-13, Chap. G). Although the possibilities of differential steric effects and phenomena due to the differing molecular geometries must be considered, it is worth noting that those complexes where the d-electron shells are completely filled (Hg(II), (Sn(IV)), lie in one group, whereas the d^6 and d^8 metals show the opposite behavior. In addition, we know that these d energy levels in the latter metals show smaller valence shell ionization potentials than do Cd or Sn; suggesting the possibility that indeed these energy levels may have to be considered in evaluating the π_{AB} terms.

We come now to the question of the polarizability which, assuming the mean triplet excitation energy approximation, is often expressed as Eq.(2.4). Most authors seem to agree that this summation, or the α^2 terms in Eq.(2.5), makes the major contribution to changes in $^1J(M,P)$. Changes in this term are thought to account for

a) negative one-bond couplings [*22,62*],

b) changes in $^1J(M,P)$ as a function of the oxidation state of the metal [*58*] and

c) the variations in this coupling as a function of the remaining ligands in the complex [*45*]. As an example consider the changes in $^1J(Pt,P)$ in the complexes $PtCl_2(PEt_3)_2$, V, and $PtCl_4(PEt_3)_2$, IV, whose metal hybrid orbitals can be thought of, if somewath naively, as dsp^2 and d^2sp^3 respectively. Considering only the change in s-character of the metal molecular orbital (dsp^2 = 25%, d^2sp^3 = 16.6%) one might anticipate an increase in $^1J(Pt,P)$, of 50%, on going from IV to V and this is observed. The agreement with this type of argument, is not always excellent (e.g. $^1J(Rh,P)$ structure (III) /$^1J(Rh,P)$ structure (II) = 1.65) although the qualitative trend is there. Not surprisingly this type of argument does not satisfactorily account for the differences in the tetrahedral $Pt(PEt_3)_4$, $^1J(Pt,P)$ = 3740 Hz and the square planar $Pt(PMe_2Ph)_4^{2+}$, $^1J(Pt,P)$ = 2342 Hz [*63*] although the metal has "25% s-character" in each case. Despite the obvious quantitative inadequacies of the s-character argument there are good reasons for believing it to be often qualitatively correct. In considering the effect on $^1J(M,P)$ of electronegative substituents on phosphorus one can consider that the phosphorus lone pair of a PF_3 or $P(OPh)_3$ ligand has more s-character relative to that of a trialkyl phosphine and that the coupling constant is corresponding larger. Support for this idea comes from known X-P-X bond angles in a variety of phosphorus ligands [*64*], as well as a recent e.s.r. study of the complexes Co(TPP)L (TPP = tetraphenylporphyrin) [*65,66*]. For L = PF_3, $P(OMe)_3$, PMe_3, PMe_2Ph, PEt_3 the % s-character in the coordinated phosphorus donor orbital is estimated to be 68, 51, 36, 28 and 27 %, respectively. Further, in the complexes $[AgP_n]X$ and $[HgP_n]X_2$ where the molecular geometries are approximately linear, trigonal and tetrahydral for n = 2, 3 and 4, respectively the values $^1J(M,P)$ nicely reflect the change in s-character of the metal-phosphorus bond (see Tables 12 and 13). The s-character-bond angle relation, commonly invoked in phosphorus chemistry, has recently been challanged by two separate sets of calculations [*67,68*], in which it has been shown that the lone pair in PH_3 has substantially more p-character (71 and 73%) than s-character (17 and 15%). This is in contrast to what one would conclude based on the H-P-H angle of $93^o50'$ alone. In retrospect the calculations demonstrate what one might have surmised based on the similarity of the lowest ionization potential in PH_3 (9.93 ev observed [*67*], 10.02 ev [*67*], 10.5 ev [*67*] calculated) and that calculated for the 3p orbital in a phosphorus atom (∿ 10.3 ev). The corresponding value for a 3s orbital is much larger. Nevertheless the same calculations [*67,68*] show that substitution of H for an electronegative group markedly increases the s-character of the lone pair (e.g. from 15% in PH_3 to 35% in PF_3). Taken in combination with calculations [*52*] which show that:

a) increase in charge on phosphorus of 0.5 electrons effects only about a 5% change in the s-expectation values and

b) that the metal expectation values are more sensitive to changes in charge than the phosphorus expectation values (and we have reason to believe that the metal $|\Psi_M(0)|^2$ is often not the major contributor to Eg. (2.4)), these data lead us to think that the differences in the $^1J(M,P)$ values with substituent electronegativity are indeed associated with changes in the polarizability and not the $|\Psi_{3s}(0)|^2_P$ term.

Assuming that the coefficients in Eq. (2.8) [or the α^2_s terms in Eq. (2.5)] make a dominant contribution to $^1J(M,P)$ implies a change in the metal-phosphorus bond lengths and this has indeed been observed. Mather et al. [*69*] have reported a correlation of $^1J(M,P)$ with the Pt-P bond length in some trialkyl phosphine complexes of Pt(II) and Pt(IV). Subsequent studies for Pt(O) aryl phosphine complexes show similar results [*70*]. Thus shorter Pt-P distances (as determined via X-ray studies) are associated with larger values of $^1J(Pt,P)$. Additionally the Pt-P separation in the complexes *trans*- $PtCl_2(PEt_3)_2$ $^1J(Pt,P)$ = 2397 Hz and *trans*- $PtCl_4(PEt_3)_2$ $^1J(Pt,P)$ = 1455 Hz are 2.298 Å and 2.393 Å respectively, further supporting the s-character argument [*70,71*]. The Cr-P distances in the complexes $Cr(CO)_5(PPh_3)$ and $Cr(CO)_5(P(OPh)_3)$ are 2.422 Å and 2.309 Å [*72*], respectively, in keeping with the rehybridization to be expected when strongly electronegative groups are bound to the phosphorus. Assuming that the W(O) complexes are analogous one can suggest that the $^1J(W,P)$ values for $W(CO)_5(PPh_3)$, 280 Hz and $W(CO)_5[P(OPh)_3]$, 411 Hz (Table 10, see Chap. G) reflect these changes in bond length. The Pt-P distances in *cis*-$PtCl_2(PEt_3)(PF_3)$ [$^1J(Pt,PEt_3)$ < $^1J(Pt,PF_3)$, l(Pt-PEt$_3$) = 2.272 Å, l(Pt-PF$_3$) = 2.141 Å] show a similar effect [*73*]. This type of correlation was not observed for the complexes $PtI_2(PMe_3)_2$ and $PtI_2(PCy_3)_2$ [*74*] where the Pt-P distances are significantly different but the $^1J(Pt,P)$ values similar ($^1J(Pt,PMe_3)$ = 2230 Hz, l(Pt,P) = 2.315 Å, $^1J(Pt,PCy_3)$ = 2288 Hz 1(Pt,P) = 2.371 Å). Despite this contradiction the possible correlation of these two methods seems worthy of further consideration. To this point we have considered $^1J(M,P)$ data solely in terms of the phosphorus ligand and the metal. Not surprisingly the remaining ligands and their geometric arrangement with respect to the phosphorus can drastically effect $^1J(M,P)$. Thus although the $^1J(Pt,P)$ for *cis*-$PtCl_2(PEt_3)_2$ is 3520 Hz, the coupling in *cis*-$PtPh_2(PEt_3)_2$ is 1705 Hz, and $^1J(W,P)$ for the two different phosphorus atoms in *cis mer*-$W(O)Cl_2(PMe_2Ph)_3$ depend strongly on whether the phosphorus is *trans* to C1 (442 Hz) or *trans* to another phosphorus (342 Hz) [*52b*]. This latter observation, e.g. $^1J(M,P)$ [*trans* to Cl] > $^1J(M,P)$ [*trans* to P] is a common one in transition metal phosphine chemistry with the ratios $^1J(P\text{-}M\text{-}Cl)/^1J(P\text{-}M\text{-}P)$ often falling between 1.3 and 1.5 [*45*]. For complexes of Pt, W, Mo, Rh and Ir the M-P bond

trans to Cl is found to be shorter than that *trans* to P. Typical examples are *cis* and *trans* $PtCl_2(PEt_3)_2$ [*75*] l(Pt-P) = 2.258 Å, l(Pt-P) = 2.314 Å and $RhCl(PPh_3)_3$ [*76*], l(Rh-P) (P *trans* to Cl) = 2.214 Å, l(Rh-P) (P *trans* to P) = 2.329 Å. The coupling constant and bond distance differences are generally attributed to the *trans* influence[6] of the ligand opposite to the M-P bond. This is an indirect effect in that the *trans* ligand exerts an influence on the metal which in turn reflects this alteration via a change in $^1J(M,P)$. The current consensus of opinion is that the M-P bond is composed of both a sigma, σ, and a pi, π, component with the former dominating. The contribution of the π component depends upon the oxidation state of the metal and the remaining ligands in the complex and synergically affects changes in the metal-phosphorus coupling [there is no provision in either Eq. 2.5 or 2.7 for a direct π contribution to $^1J(M,P)$]. The increase in $^1J(W,P)$ with presumed decreasing W-P distance observed when comparing $W(CO)_5(P(OPh)_3)$ (411 Hz) with *trans*-$W(CO)_4(P(OPh)_3)_2$ (480 Hz) has been rationalized [*40,77*] using a π bonding argument. It is reasoned that, since C≡O should compete more effectively than phosphorus for metal d-orbital electrons, the coupling will be smaller when the phosphorus is *trans* to the better π bonding ligand, however, this argument has been questioned [*88*]. The subject of the *trans* influence has been reviewed [*78*] and a number of physical methods including X-ray, IR, ^1H and ^{13}C NMR, are commonly employed in its study. The applications of ^{31}P NMR in this area are surprisingly few with the order of *trans* influence from measurements of $^1J(Pt,P)$ reported to be:
$MePh_2Si$ > Ph > Me >> R_3P > $(PhO)_3P$, CN > Et_3As > NO_2 > amine, *N*CO, *N*CS > Cl, Br, I > ONO_2 [*46,57*].

The effects on $^1J(M,P)$ which result when the inter-ligand angles are constrained via chelation are quite significant. In the Pt(O) complexes Pt(triphos)L, L = PPh_3, $P(pTol)_3$, $PMePh_2$, $P(OCH_2)_3CCH_3$, PF_2NMe_2, $P(OPh)_3$, PF_3, $^1J(Pt,P_L)$ values between 5370 and 9150 Hz were observed [*79*]. The couplings in the tridentate ranged from 2837 to 3098 Hz. Since the triphosphine ligand restricts the P-Pt-P angles to slightly < 90° the unusually large $^1J(Pt,P_L)$ values have been attributed to increased s-character in this Pt-P bond. The values $^1J(M,P)$ in W [*80*] and Pt [*81*] complexes containing the bidentate ligands dppm, dppe and dppp have been shown to depend on the number of CH_2 groups connecting the phosphorus atoms and we shall return to this point in connection with the discussion on phosphorus chemical shifts.

---

6. The *trans* influence is defined as the extent to which a ligand weakens a bond *trans* to it in the equilibrium state [*58*].

The system HgX_2P_2 has been extensively studied [82] using both ^{31}P NMR and x-ray methods and is interesting in that $^1J(Hg,P)$ can be correlated to:

a) the Hg-P bond distance
b) the X-Hg-X and P-Hg-P bond angles
c) the $\langle 6s(Hg) | 3s(P) \rangle$ overlap integral
d) the ionization potential of the X ligand (see Fig. 11) and
e) a nucleophilicity scale, E_N for the X ligand.

A summation of the ^{31}P NMR and x-ray structural data for these mercury complexes is shown in Table 14 (see Chap. G).

In contrast to other systems $^1J(Hg,P)$ for the HgX_2P_2 complexes is extremely sensitive to X with values observed between 576 and 6,003 Hz. This is the largest range yet observed for a metal-phosphine coupling with the larger value of the coupling observed for weaker donor X ligands. The structural data reveal that, in addition to the

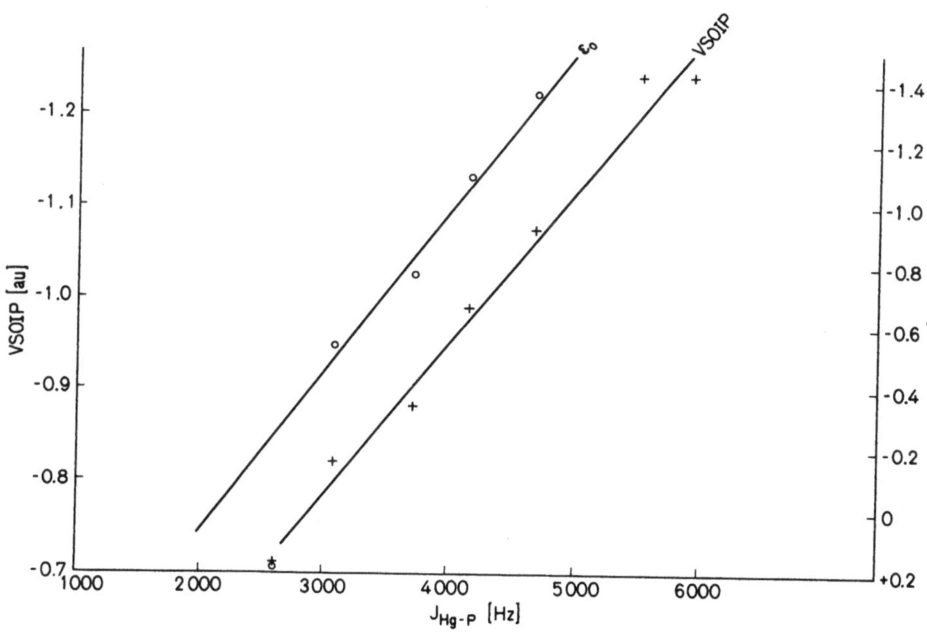

Fig. 11. Plot of the Valence State Orbital Ionization Potential, VSOIP (Vertical left) and the electrode potential for the oxidative dimerization of the anion, X. (vertical right) against $^1J(Hg,P)$.

expected decrease in 1(Hg,P) with increasing coupling constant, the P-Hg-P angle opens (131° in $Hg(NO_3)_2(PPh_3)_2$) and the X-Hg-X angle closes with increasing coupling constant. Constricting the P-Hg-P angle while simultaneously introducing a strongly coordinated X ligand leads to a markedly reduced metal phosphorus coupling. The effect of holding X constant and changing the phosphine roughly follows the electronegativity effect referred to previously. Since Hg(II) is a d^{10}, doubly charged system the likelihood of metal d-orbital participation is small and one need not be concerned about synergic π interactions. Initial Extended Hückel calculations reveal the expected dependence of the one bond mercury-phosphorus coupling on π_{AB} and have confirmed a significant dependence of π_{AB} on the inter-ligand angles [*82*]. The large values of J(Hg,P) and the sensitivity of this coupling to structural changes makes it an attractive candidate for further study.

In summation, the generally accepted theory for one bond couplings seems valid for $^1J(M,P)$. The polarizability, π_{AB}, seems to make the dominant contribution to the Fermi contact term and within this framework, one can rationlize the existing empiricisms and correlate metal-phosphorus coupling constants with other physical methods.

II Two-Bond Phosphorus-Phosphorus Coupling Constants

Amongst the most widely studied phosphine complexes are those containing two and three phosphorus atoms. Such molecules will give rise to ^1H, ^{13}C and, when the phosphorus atoms are chemically non-equivalent, ^{31}P NMR spectra in which the magnitude of the coupling $^2J(P,P)$ will strongly affect the spectral appearance. In favorable cases the spectra will be first order and directly deliver structural information. Fig. 12 shows the $^{31}P\{^1H\}$ spectrum of $PtCl(SnCl_3)$(DIOP) from which one can deduce that the molecule contains two different types of phosphorus and thus that the molecule probably has a *cis* geometry. The magnitude of this two-bond coupling is known to be dependent on whether the atoms are *cis* or *trans* to one another thereby providing an additional structural tool. This type of empiricism has been found for a variety of $^2J(P\text{-}M\text{-}X)$, values where X = ^1H [*83*], ^{13}C [*84*], ^{15}N [*85,86*], $^{107,109}Sn$ [*87*], ^{199}Hg [*88*], and seems to be a general phe-

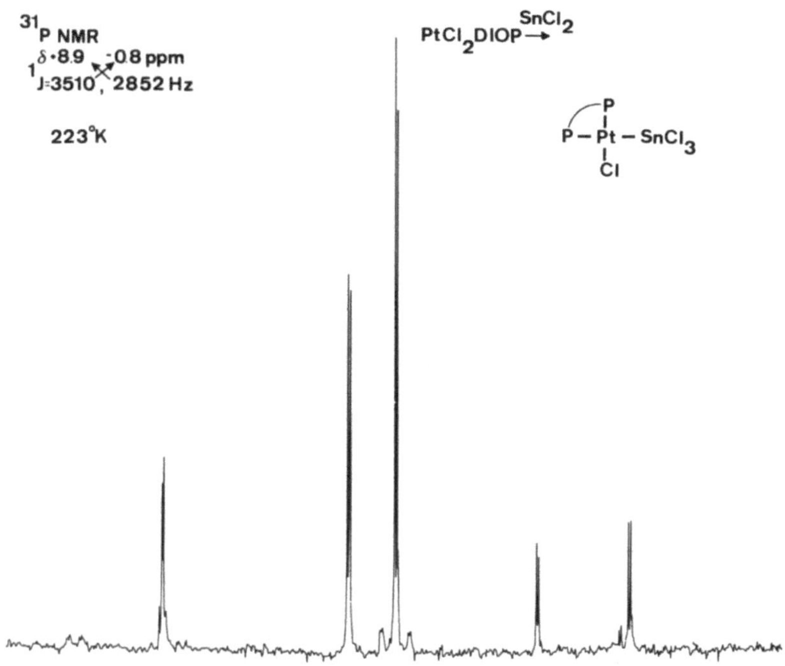

Fig. 12. $^{31}P\{^1H\}$ NMR spectrum of $PtCl_2$(DIOP)

nomenon. Examples of this type of coupling are shown in Table 17 (Chap. G).

Since bis phosphine derivatives are found for a large number of metals it is worth considering the value $^2J(P,P)$ in some detail. These data are generally obtained via:

a) direct observation of the ^{31}P NMR spectrum. This is useful when the complex contains chemically non-equivalent phosphorus ligands thus affording simple or readily analysable NMR spectra or
b) indirect observation using heteronuclear double resonance methods or
c) direct observation of a second order ^1H, ^{13}C or ^{19}F NMR spectrum. The detailed analysis of such a second order spectrum e.g. of the type $X_nAA'X'_n$, A = ^{31}P, X = measured nucleus has been reported [*90*] and a schematic diagram of a commonly observed solution (n = 6) is shown in Fig. 13. For those cases where the weak outer resonances can be observed $^2J(P,P)$ can be measured directly. When these weak resonances are not visible (their intensities decrease as $^2J(P,P)$ increases) one can estimate the homonuclear two bond

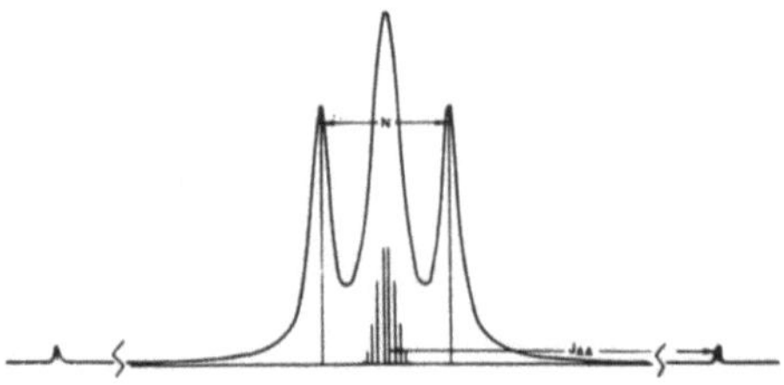

Fig. 13. Schematic showing the solution for the X part of an $X_6AA'X'_6$ spin system [*92*]

coupling from the line width of the central resonance; however, this approach has been criticized and methods a) and b) are thought to be preferable to c)

The theory invoked to explain changes in $^2J(P,P)$ represents an empirical extension of the LCAO-MO treatment described in the previous section. The Fermi contact term is assumed dominant and this expression takes the form of Eq. (2.3).

$$^2J(P,P) \propto |\Psi_{3S}(0)|^2_P |\Psi_{3S}(0)|^2_{P'} \sum_i^{occ} \sum_j^{unocc} c_{iP}c_{jP}c_{iP'}c_{jP'}/(E_i-E_j)$$

Using this relation we can attempt to rationalize those factors which have been shown empirically to affect this coupling. These are:

a) the geometric position of the two phosphorus atoms. Under this heading we include the dependence of both the magnitude ($^2J(P,P)_{trans} > {}^2J(P,P)_{cis}$) and sign ($^2J(P,P)_{trans}$ often > 0, $^2J(P,P)_{cis} < 0$) on the complex stereochemistry.
b) the metal
c) the remaining ligands
d) the types of substituent bound to phosphorus.

An explanation for these observations is available if one assumes that the singlet-triplet excitation plays a dominant role in determining $^2J(P,P)$. In order to visualize how changes in this term affect the two bond coupling we require an exact knowledge of the molecular orbital energy level diagram. This is generally not available; however, assuming that Fig. 14 and 15 are reasonable for some simple octahedral complexes one can proceed further. We will assume that an excitation from one orbital to another of the same symmetry contributes a negative sign to $^2J(P,P)$ while a transition involving orbitals

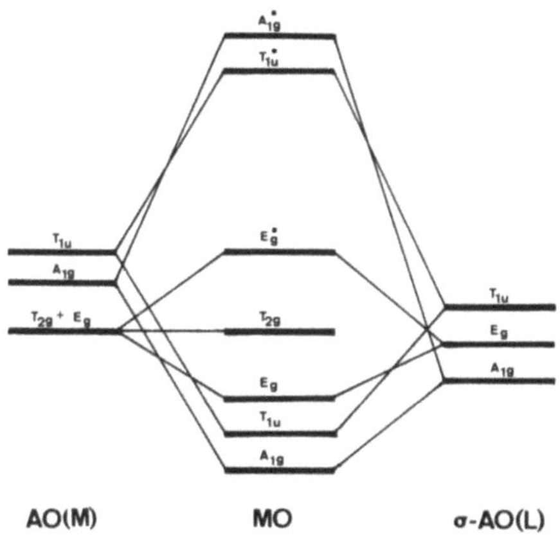

Fig. 14. The MO scheme for the octahedral molecule "MP_6"

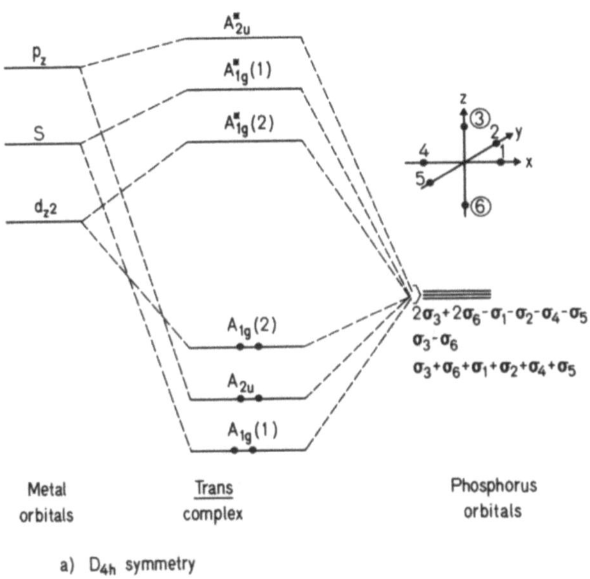

Fig. 15. The MO scheme for a molecule of D_{4h} symmetry [*92*]

of opposite symmetry produces a positive contribution to this coupling. Further, we will consider only σ ligand orbitals for which there are non-vanishing s-orbital contributions for both phosphorus nuclei. For the hypothetical MP_6 the lowest energy transition is of the type $E_g \rightarrow E_g{}^*$ and should contribute negatively to $^2J(P,P)$. (Note that there is no contribution to $^2J(P,P)_{cis}$ from t_{1u} ligand orbitals since these orbitals cannot combine with both phosphorus orbitals to give a non-vanishing s-orbital contribution, see Appendix). If one neglects differences in the terms $c_i c_j$ and assumes that the lowest energy transition is most important then one can account for the observation of negative $^2J(P,P)_{cis}$ values. Consider now the proposed partial diagram for *trans*-$M(CO)_4P_2$ [*91*] for the case where the energy of the phosphorus lone pair is less than that of the carbon lone pair. In this case the transition $A_{2u} \rightarrow A_{1g}^*$ is lower in energy than the $A_{1g} \rightarrow A_{1g}^*$ and will contribute positively. Obviously, this type of argument is qualitative in the extreme and cannot stand alone[7]; however, it does account for the observation that most *cis* and *trans* complexes have negative and positive values of $^2J(P,P)$ respectively [*92*].

It is generally observed that as one descends in the Periodic Table $^2J(P,P)$ increases algebraically. Thus for *trans*-$M(CO)_4(P(OMe)_3)_2$, M = Cr, Mo, the values are -15 and +162 Hz respectively, whereas for *trans*-$M(CO)_4(PBu_3^n)(PPh_3)$, M = Cr, Mo, W, these couplings are 25, 50 and 65 Hz, respectively. Returning to Fig. 14, and remembering that the d-electrons decrease in energy as one goes down a column, one can imagine that the E_g level falls relative to the T_{1u} and A_{1g} levels, thus decreasing the negative contribution, while increasing the positive contribution to $^2J(P,P)$. This idea is also consistent with the observation that $^2J(P,P)_{cis}$ and $^2J(P,P)_{trans}$ values for metal complexes in the first transition series can be of similar magnitude whereas in the second and third series the *trans* coupling is usually much larger[8].

Changes in the ΔE term have also been invoked to explain the changes in $^2J(P,P)$ in both tetrahedral and square planar complexes with the appropriate diagrams shown in Fig. 16 and 17, respectively. In the former case the transitions stemming from the T_2 and T_2' orbitals reaching to the A_g^* and T_2^* orbitals are considered to be of primary importance. As these last two levels may have similar energies and since the $T_2 \rightarrow T_2^*$ and $T_2 \rightarrow A_1^*$ transitions will result in coupling of opposite sign, small values of $^2J(P,P)$ are predicted and, for complexes of the type NiP_4 [*93*], are observed. In the planar complexes, the $B_{3u} \rightarrow A_g^*(2)$ transition is thought to be dominant [*91*]. Since the B_{3u} level is derived from a metal p-orbital and the $A_g^*(2)$

7. Extension of this type of argument to *cis* $M(CO)_4P_2$ gives a more complicated somewhat ambiguous result. A more exact knowledge of the orbital coefficients and relative energies is necessary for a meaningful discussion.

8. In considering the octahedral complexes ML_4P_2 we have neglected the relative energies of the phosphorus and L ligands. This has also been suggested as important when seeking to explain why $^2J(P,P)_{cis}$ can be larger than $^2J(P,P)_{trans}$ for some first row metal complexes [*92*].

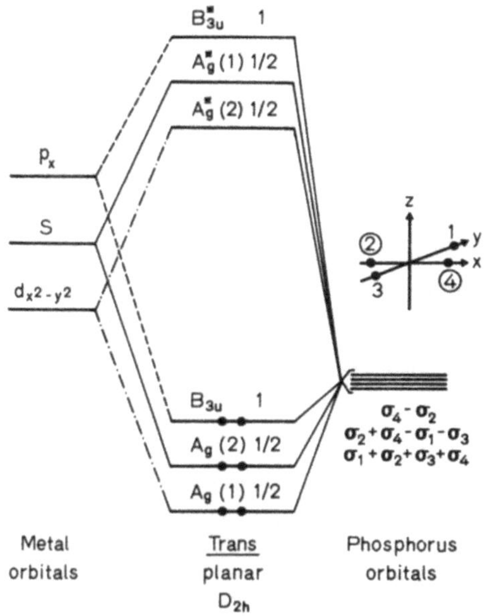

Fig. 16. The MO scheme for a *trans* planar complex with symmetry D_{2h} [*92*]

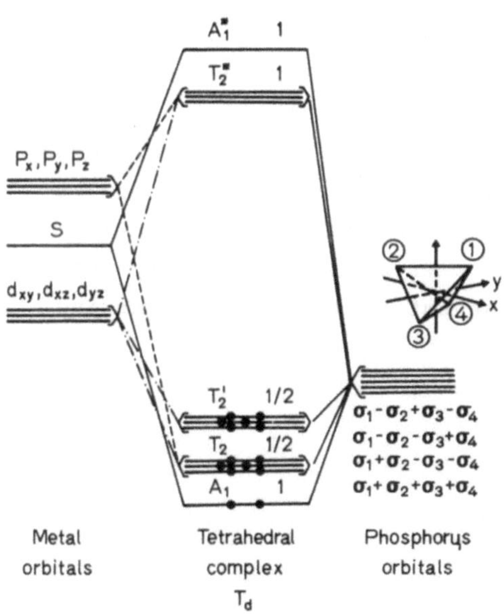

Fig. 17. The MO scheme for a tetrahedral complex with symmetry, T_d [*92*]

contains metal d-orbital character, an increase in the separation of the nd- and (n+1)p-levels should result in a decrease in the magnitude of $^2J(P,P)$. This d-p separation increases a) as the oxidation state of the metal increases b) in the order Ni < Pt < Pd and Ir < Rh and c) in the order Ru < Rh < Pd [*94*]. From Tables 18 -21 (Chap. G) we can see that $^2J(P,P)$ does indeed qualitatively fulfill this expectation.

The effect of other complexed ligands on $^2J(P,P)$ seems to be irregular and as relatively few data are available the point remains open; however, this coupling has been found to increase in the order I < Br < Cl [*92*].

Much more pronounced is the effect on $^2J(P,P)$ of the other groups bound to phosphorus. In particular, there are several reports relating the electronegativity of R to $^2J(P,P)$ in complexes containing PR_3 ligands. There are two ways in which this can occur within the framework of our theory:

a) $|\Psi_{3S}(0)_P|^2$ can increase with increasing partial positive charge on phosphorus. Thus the s-density increases as the electronegativity of R increases and
b) the eigenvectors of the phosphorus s-orbitals change as a function of the electronegativity of R thus changing the polarizability term.

Point b) can affect a change in both sign and magnitude whereas a) should affect only the magnitude of the coupling. For some complexes of Cr, Mo and W there seems to be a relatively smooth increase in the algebraic value of both $^2J(P,P)_{trans}$ and $^2J(P,P)_{cis}$ as the electronegativity of R increases (see Fig. 18) suggesting that point a) may be important [*85*], whereas in some nickel(0) derivatives the correlation is not clear and both a) and b) may be contributing. Obviously the theory is not quite up to the experiment but it is certainly making strides in the correct direction.

III One-Bond Metal-Phosphorus Coupling Constants – A Survey

The previous paragraphs have concerned themselves with the qualitative aspects of the theory of one-and two-bond phosphorus couplings involving metals. In a number of instances we have cited specific if somewhat selective literature to support these ideas. The section which follows provides a more comprehensive survey of the types of molecules for which one-bond metal-phosphorus couplings have been reported.

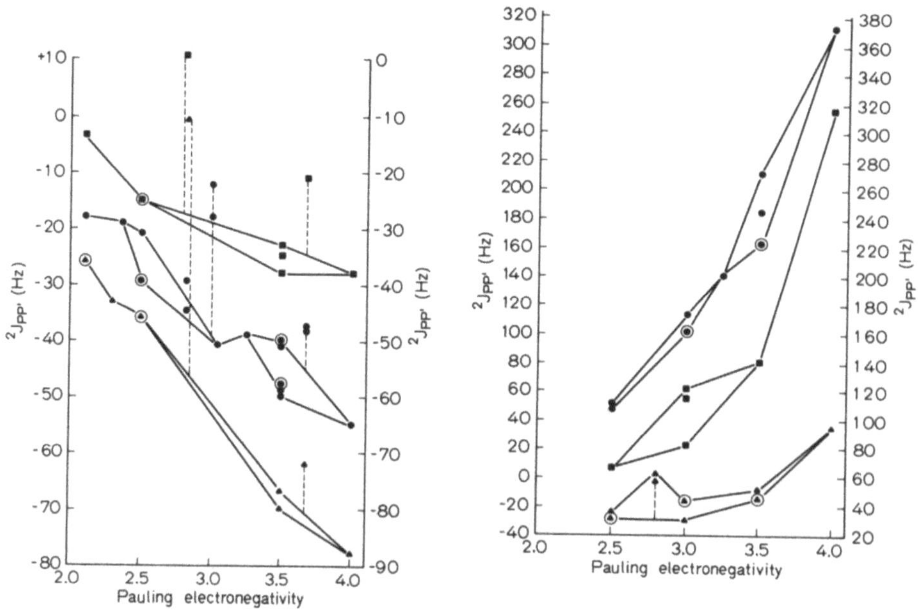

Fig. 18. Plots of $^2J(P,P)$ vs the Pauling Electronegativity for some (left) *cis*-M $(CO)_4P_2$ and (right) *trans*-M$(CO)_4P_2$ complexes [*92*] M = Cr (Δ), Mo (o) or W (□)

Group Ib

In the group, Cu, Ag, Au most emphasis has been placed on Ag since this metal possess two spin I = 1/2 isotopes, ^{107}Ag and ^{109}Ag, in almost equal proportions. The difference in coupling to phosphorus stemming from the different silver isotopes is readily observable and is directly proportional to the ratio of their gyromagnetic constants. $^1J(Ag,P)$ values have been reported for the complexes $Ag(PBu_3{}^n)_nBF_4$, n = 1-4 [*95*], $Ag(P(pTol)_3)_nX$ n = 2-4 [*96*], $Ag(P(OEt)_3)_nX$, n = 2-4 [*102*] and Ag(PP)X, where PP = 2,11-Bis(diphenylphosphinomethyl)benzo [c]phenanthrene and X is a variety of anionic ligands [*97*]. The values $^1J(Ag,P)$ for the complexes $[AgP_2]_2Ni(S_2C_2(CN)_2)_2$, P = PPh_3, $PEtPh_2$, PEt_2Ph, PEt_3 have recently appeared [*98*]. These studies have primarily concerned themselves with the dependence of $^1J(Ag,P)$ on the coordination number of the metal and the nature of the remaining ligands within the coordination sphere. Typically, $^1J(^{107}Ag,P)$ values for phosphine complexes fall between 200 and 400 Hz. A sampling of data is shown in Table 15 (Chap. G).

$^{1}J(^{63}Cu,P)$ values for the phosphite complexes $Cu(P(OMe)_3)_4^+$ [*27*] and $Cu(P(OEt)_3)_4^+$ [*27*] have also been reported. ^{63}Cu is a quadrupolar nucleus with $I = 3/2$; however, the symmetry of the molecules is favourable in these cases.

Group IIb

In the triad Zn, Cd, Hg, phosphine complexes of the latter metal have attracted the most attention. ^{199}Hg has $I = 1/2$ and a natural abundance of 16.9%. A typical $^{31}P\{^{1}H\}$ spectrum shows a main-band to satellite ratio of ≈ 10:1. The molecules $HgX_2(PBu_3^n)_2$, X = a variety of anionic ligands, have been investigated by four different groups [*49-51,99*]. When phosphite replaces phosphine the one-bond coupling increases markedly [*49*]. $^{1}J(Hg,P)$ values for the dimeric species $Hg_2Cl_4P_2$ have been measured and are generally significantly larger than those for the corresponding mononuclear complexes ($^{1}J(Hg,P)$ $HgCl_2(PBu_2^nPh)_2$ = 5035 Hz; $^{1}J(Hg,P)$ $Hg_2Cl_4(PBu_2^nPh)_2$ = 7514 Hz [*50*]). The effect of altering the phosphine [*50*] and X groups [*49,50*] on both $^{1}J(Hg,P)$ and the molecular structure [*82*] has been described. The linear complexes $HgX(PO(OEt)_2)$, X = OAc, Cl, Br, I, $PO(OEt)_2$ show $^{1}J(Hg,P)$ values which vary from 7.50 - 12.97 KHz [*100*]. The one-bond coupling for the complexes $Hg(PMe_3)_n^{2+}$, n = 2-4 decreases with increasing n (5505, 3580 and 2255 Hz for n = 2, 3, 4 respectively) thus clearly demonstrating the importance of the coordination number for this type of coupling [*51*]. A sampling of data is shown in Table 13 (Chap. G).

There have been relatively few studies describing the coupling of cadmium to phosphorus although there are two $I = 1/2$ isotopes of this metal, ^{111}Cd and ^{113}Cd, in approximately equal proportions (12.9 and 12.3% respectively). These two isotopic one-bond couplings are readily distinguishable and have been observed in the complexes CdI_2P_2, P = PMe_2Ph, $PMePh_2$, PEt_3, PEt_2Ph, $PEtPh_2$ [*101*] and the dimeric species $Cd_2X_4(PBu_3^t)_2$, X = Cl, Br, I, SCN [*102*]. For comparable molecules the dimeric species show a much larger one-bond coupling in analogy with mercury complexes. Typical values for $^{1}J(^{113}Cd,P)$ in the monomeric and dimeric species are ∿ 1200 Hz and ∿ 2000 Hz, respectively.

Group Va

There are two reports [*23,24*] of vanadium phosphorus one-bond couplings. (^{51}V, I = 7/2, 99.8% abundant) and both of these resulted from the direct measurement of the metal NMR spectrum. Data are given for 49 mono- and disubstituted phosphine derivatives of η^5-CpV(CO)$_4$ and V(CO)$_6^-$. The $^1J(V,P)$ values decrease in the order PBu$_3^i$ > PF$_3$ > P(OR)$_3$ > P(NR$_2$)$_3$ > PR$_3$ > PH$_3$. In the complexes η^5-CpV(CO)$_2$(PR$_3$)$_2$, $^1J(V,P)_{trans}$ > $^1J(V,P)_{cis}$ by a factor of ca. 1.25. Typical values for the V(CO)$_5$(PR$_3$)$^-$ complexes lie between 170 and 730 Hz.

The molecule TaH(CO)$_2$(Me$_2$PCH$_2$CH$_2$PMe$_2$)$_2$ has been investigated via ^{31}P and ^1H NMR methods [*103*].

Group VIa

For the Cr, Mo, W triad, the metal carbonyl-phosphine complexes of tungsten (^{183}W I = 1/2, natural abundance = 14.3%) have received attention since a wide variety of these complexes is easily synthesized. There have been several groups who have reported $^1J(W,P)$ values for the complexes W(CO)$_5$(PR$_3$) [*40,43,59,104,105*] and *cis* and *trans* -W(CO)$_4$(PR$_3$)$_2$ [*77,106,107*]. Additionally, there are studies describing the ^{31}P NMR characteristics of the complexes *fac* -W(CO)$_3$(PR$_3$)$_3$ [*77*], and [η_5CpW(CO)$_2$LX], where L = a phosphine or phosphite ligand and X = Me$_3$Sn or Me$_3$Pb [*108 a,b*]. $^1J(W,P)$ in the pentacarbonyl phosphine complexes has a positive sign. Two separate linear relationships have been found in the W(CO)$_5$PR$_3$ complexes connecting the Sanderson Electronegativity of the substituents on phosphorus and $^1J(W,P)$ [*43*]. The values of $^1J(W,P)$ in this series span the range 143 - 485 Hz [*59,104*]. The extent to which metal-phosphorus d→p π back bonding influences the magnitudes of the one-bond coupling in the *cis* and *trans* W(CO)$_4$(PR$_3$)$_2$ derivatives, structures VII and VIII, remains under discussion [*77,106*].

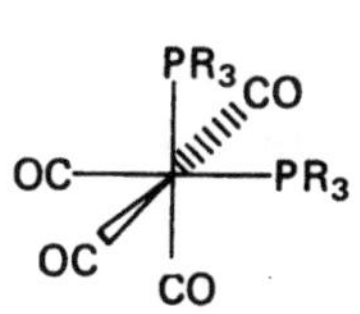

(VII)

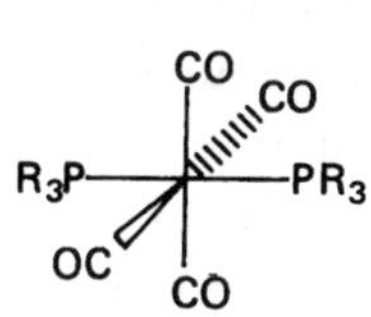

(VIII)

In addition to the studies involving monodentate phosphines there have been a relatively large number of reports involving bidentate phosphine ligands and some involving tri- and tetradentate systems. The ligands $Ph_2PC{\equiv}CPPh_2$ [*109*], $Ph_2PCH_2PPh_2$ [*80*], $Ph_2PCH_2CH_2PPh_2$ [*80*] $Ph_2PCH_2CH_2CH_2PPh_2$ [*80*], $Ph_2PCH_2P(OPh)_2$ [*80*], $Ph_2PCH_2CH_2P(OPh)_2$ [*80*], $Ph_2P(S)CH_2PR^1R^2$ [*80*] and $Ph_2PCH_2PR^1R^2$ [*80*] have all been employed. In some instances the phosphorus ligands function as bidentates affording monomeric products, in other cases only one of the two phosphorus atoms coordinates leaving an uncomplexed phosphorus atom [*109*]. For this latter class of complexes there is the possibility that the free phosphine will bridge a second metal and this has been found as well [*110*]. For the potentially tetradentate ligand $P(o\text{-}PPh_2C_6H_4)_3$, QP, similar possibilities exist and have been observed [*111*]. In these latter molecules the magnitudes of the values $^1J(W,P)$ have been found to be useful structural aids [*112*]. An extensive amount of $^2J(P,P)$ information is available for the bidentate ligands containing chemically non-equivalent phosphorus atoms as well as for some "mixed" ligand complexes [*80*]. A selection of $^1J(W,P)$ values is shown in Table 10 (Chap. G).

$^1J(^{95,97}Mo,P)$ values for complexes of the type $Mo(CO)_5L$, $L = PF_3$, $P(OPh)_3$, $P(OCH_2)_3CEt$, $P(OMe)_3$, $P(OEt)_3$, $PPh(OMe)_2$, $PPh_2(OMe)$, PPh_3, $P(NMe_2)_3$ and $PPhMe_2$ have been measured with values ranging from 133 - 284 Hz [*28*].

Group VIII

There is little to be said with regard to the Fe, Ru and Os triad, although $^1J(^{57}Fe,P)$ values in the complexes $Fe(CO)_4(PEt_nPh_{3-n})$ (25.9, 26.5 and 27.4 Hz for n = 3, 2 and 1, respectively [*25*]) have been described as have two one-bond osmium-phosphorus values for the complexes $OsH_4(PEt_2Ph)_3$ and *cis*-$OsCl_2(CO)_2(PBu^tPr_2^n)_2$ $^1J(^{187}Os,P)$ = 166 and 149.8 Hz, respectively [*44*]; however, the Co, Rh, Ir and Ni, Pd, Pt groups have attracted considerable attention with the largest volume of coupling constant data centering around rhodium and platinum. Without doubt ^{195}Pt (I = 1/2, natural abundance = 33.7%) leads the way with something approaching one hundred reports in which the values $^1J(Pt,P)$ are given, with ^{103}Rh (I = 1/2, natural abundance = 100%) running a respectable second. It is convenient to subdivide the rhodium studies into three sections:

a) 4-coordinate planar Rh(I) complexes
b) 5-coordinate Rh(I) and Rh(III) complexes
c) 6-coordinate octahedral Rh(III) complexes

Most of the work involving this metal concerns itself with the chemistry of these three classes of compounds and the transformations from one to another. Often the ^{31}P NMR is incidental and is included primarily for the sake of completion. A noteworthy exception is a recent study in which the reduced one-bond rhodium-phosphorus couplings for complexes of Rh(I) and Rh(III) were shown to be positive [*113*]. Several groups have observed that $^{1}J(Rh,P)$ for a phosphorus *trans* to a halogen is larger than $^{1}J(Rh,P)$ for two phosphorus atoms *trans* to one another [*30,31*]. This is the case for *both* Rh(I) and Rh(III) complexes [*114,115*]. The well known Wilkinsons catalyst, $RhCl(PPh_3)_3$, IX, is a typical example:
$^{1}J(Rh,P_A)$ = 192 Hz, $^{1}J(Rh,P_B)$ = 146 Hz [*116a*] (for $RhCl(P(p\text{-}Tol)_3)_3$, $^{1}J(Rh,P_A)$ = 189 Hz, $^{1}J(Rh,P_B)$ = 143 Hz [*116b*]. There has been considerable interest in the complexes *trans*-$RhCl(CO)P_2$, P_2 = a variety of

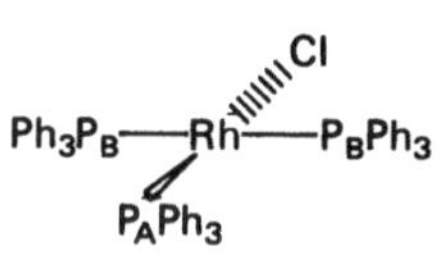
(IX)

monodentate and bidentate phosphine ligands. In these molecules $^{1}J(Rh,P)$ falls in the range 117 - 129 Hz with higher values found for phosphorus ligands containing strongly electronegative groups [*29,117*]. Small changes are observed when the chloride ion is substituted (e.g. for *trans*-$RhX(CO)(PPh_3)_2$, $^{1}J(Rh,P)$ = 129, 126 and 132 Hz for Cl, Br and OAc respectively [*117*]. This coupling can be somewhat larger as in structures X and XI, which result from the demethylation of RhCl$(CO)(PBu^t_2(2\text{-}MeOC_6H_4))_2$. Both of these conformers are distinguishable on the NMR time scale at RT [*118,119*]. We will return to this type of molecule in a later section.

A number of 5-coordinate rhodium phosphine complexes have been reported and these may have either trigonal bipyramidal, XII, or square pyramidal, XIII, structures. For a series of $RhL_5^+X^-$ complexes, where L is a phosphite, $P(OR)_3$, and X a non-coordinating anion, low temperature $^{31}P\{^{1}H\}$ studies gave spectra of the type A_2B_3X (X = ^{103}Rh) thus, proving structure XII to be correct [*120*]. Interestingly, $^{1}J(Rh,P)$

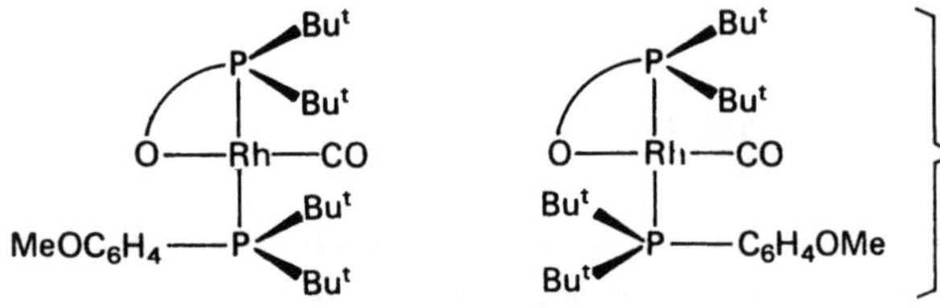

(X) (XI)

$^1J(Rh, P)$ = 131, 139 Hz

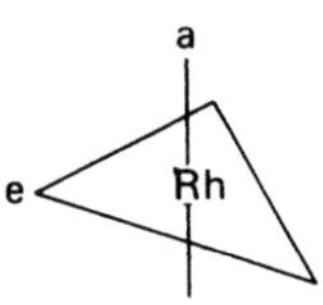

(XII)

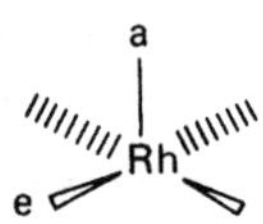

(XIII)

for equatorial phosphite is larger than for the corresponding axial phosphite (for $[Rh(P(OMe)_3)_5]BPh_4$, $^1J(Rh,P_e$ = 206 Hz, $^1J(Rh,P_a)$ = 143 Hz) [*120*]. A similar dependence of $^1J(Rh,P)$ has been noticed in the trigonal bipyramidal complex, $RhCl(Ph_2POCH_2CH{=}CH_2)_2$, XIV, and other related complexes in which one phosphorus occupies an equatorial and the other an axial position [*121*]. A trigonal bipyramidal arrangement has been suggested for $RhCl_2(PhCH_2CH_2CO)(PPh_3)_2$, XV, with the two

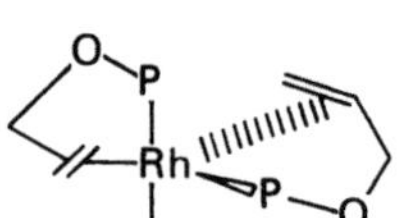

(XIV)

$^1J(Rh, P_e)$ = 146 Hz
$^1J(Rh, P_a)$ = 116 Hz

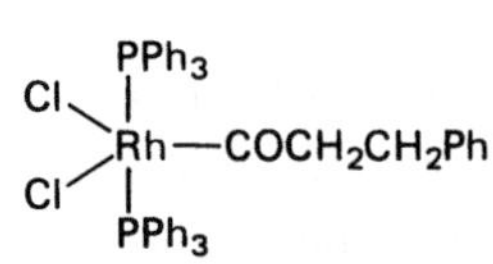

(XV)

phosphines occupying apical positions ($^1J(Rh,P)$ = 108 Hz [*122*]. The trigonal bipyramidal arrangement of the β-phenacyl derivative, XV, is supported by a preliminary x-ray structure determination which shows the two phosphorus ligands to be in apical positions [*122*]. The marked difference in $^1J(Rh,P)$ in the two square pyramidal complexes XVI and XVII [*123*] suggests that the geometric dependence of $^1J(Rh,P)$ found for four- and six-coordinate complexes is also valid for the case of 5-coordination. Given the similar nature of the ^{31}P parameters for XV, δ = 23.2 ppm $^1J(Rh,P)$ = 108 Hz, and XVII δ = 23.6 ppm, $^1J(Rh,P)$ = 108 Hz the proposition that these have different structures seems dubious although the complex RhCl(NO)(tpp), which is known to have a square pyramidal structure, shows $^{1}J(Rh,P)$ = 106 Hz for the *trans*-P atoms [*124*]. The complex $RhCl_2(CH_2{=}CHCH_2CO)(PMe_2(o\text{-}OCH_3C_6H_4))_2$ shows equivalent phosphines with $^{1}J(Rh,P)$ = 146.5 Hz and is thought not to have apical phosphine atoms [*119*]. Its ^{1}H spectrum suggests that the two phosphines are not *trans*. Data for $Rh(NO)(PPh_3)_3$ and $Rh(NO)Cl_2(PPh_3)_2$ have been reported [*125*].

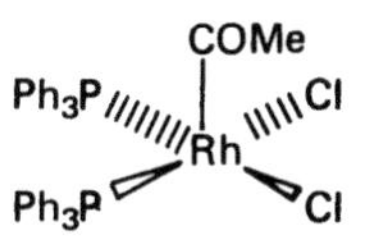

$^{1}J(Rh, P)$ = 145 Hz

(XVI)

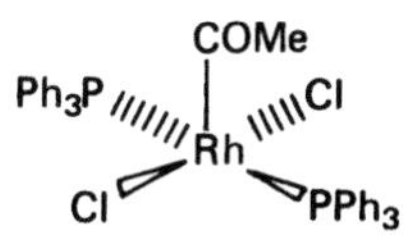

$^{1}J(Rh, P)$ = 108 Hz

(XVII)

In the octahedral trisphosphine Rh(III) complexes, RhX_3P_3, both the facial, XVIII, and the meridional, XIX, complexes can be isolated and several studies have reported their ^{31}P NMR characteristics [*30, 31,114,115*]. There is relatively little change in $^{1}J(Rh,P)$ as a function of the phosphine ligand.

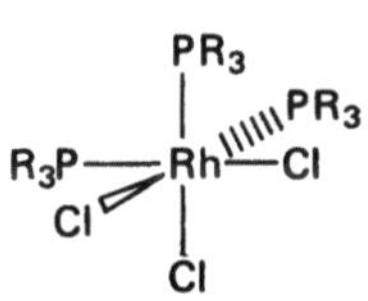

$^{1}J(Rh, P) \sim$ 103–115 Hz

(XVIII)

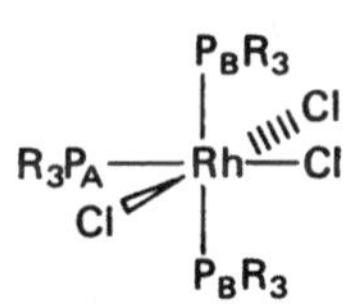

$^{1}J(Rh, P_A) \sim$ 103–120 Hz $^{1}J(Rh, P_B) \sim$ 84–86 Hz

(XIX)

The interesting structures, XX and XXI, containing bridging halogens [*126*] and hydrides [*127*], respectively show typical one-bond couplings.

The "mixed" bimetallic complex $RhCl(PBu^{t}Pr_2)_2(\mu\text{-}Cl)_2PtCl(PBu^{t}Pr_2)$ in which the phosphines are axial and *trans* shows $^{1}J(Rh,P)$ = 95.6 Hz [*128*]. Many of the molecules mentioned represent new and interesting molecular types, e.g. XXII, which contains a rhodium-mercury bond, and we shall return to these at a later point.

The complexes for which one-bond platinum-phosphorus couplings (^{195}Pt, I = 1/2 natural abundance = 33.7%) have been tabulated are

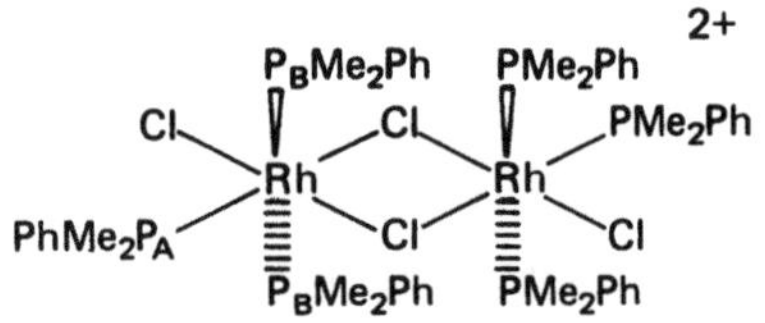

$^{1}J(Rh, P_A)$ = 119 Hz, $^{1}J(Rh, P_B)$ = 83 Hz

(XX)

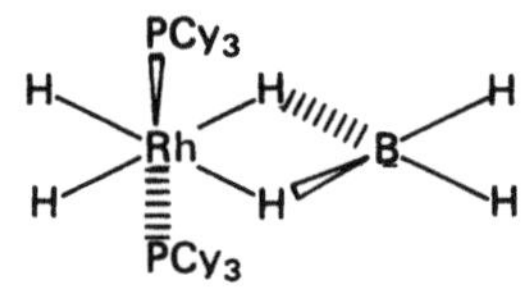

$^{1}J(Rh, P)$ = 108 Hz

(XXI)

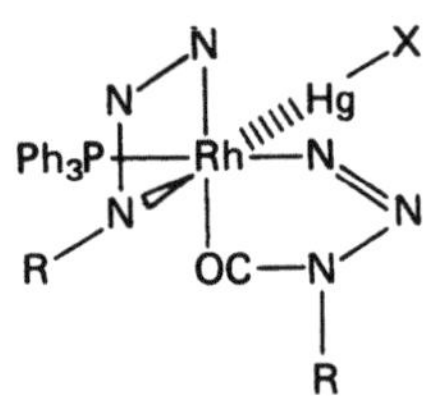

$^{1}J(Rh, P)$ = 133 Hz

(XXII)

conveniently divided into two classes: those in which the NMR spectroscopy features prominently and those in which it is incidental.

In the first category one finds that emphasis is generally placed on the relation between the chemical environment of the metal and $^{1}J(Pt,P)$. In particular, there is a dependence of $^{1}J(Pt,P)$ on a) the oxidation state of the metal [*58*], b) the *trans* influence of the remaining ligands in the complex [*46, 57, 58, 129, 130*] c) the size of the chelate ring (for polydentate ligands) [*131-133*] and d) the type of phosphorus ligand [*134*]. In a similar fashion to $^{1}J(W,P)$ and $^{1}J(Rh,P)$, $^{1}J(Pt,P)$ has a positive reduced coupling [*135, 136*] and a small dependence on solvent [*137*]. Taking points a-d sequentially, one observes large differences in $^{1}J(Pt,P)$ as a function of the metal oxidation state with smaller values for Pt(IV) and larger values for Pt(II) and Pt(0). The typical range for this type of coupling is 1,000 - 4,500 Hz (e.g $^{1}J(Pt,P)$ for *fac*-$PtMe_3I$(dppe) = 1100 Hz, [*136*], whereas $^{1}J(Pt,P)$ for the chloride bridged $Pt_2Cl_4(PMe_2Ph)_2$ = 3,900 Hz [*136*]). Since we have discussed the question of metal oxidation state earlier the reader is referred to the appropriate references and Tables 3 - 9 (Chap. G) for further detail. The dependence of $^{1}J(Pt,P)$ on the ligand *trans* to phosphorus is a recurring theme. Differences of slightly more than a factor of two can be induced by varying the *trans* ligand from something with a small *trans* influence (e.g. a halogen or oxygen ligand) to one with a much larger *trans* influence (e.g. another phosphorus or an alkyl group [*129*]). Thus the two different one-bond couplings in *cis*-$PtClMe(PEt_3)_2$ are 4179 Hz (P *trans* to Cl) and 1719 Hz P *trans* to Me [*138*].

The effect of chelate ring size on $^1J(Pt,P)$ is illustrated by structures XXIII - XXV. [*131*], with the four membered ring showing the most pronounced effect. The nature of the phosphorus ligand,

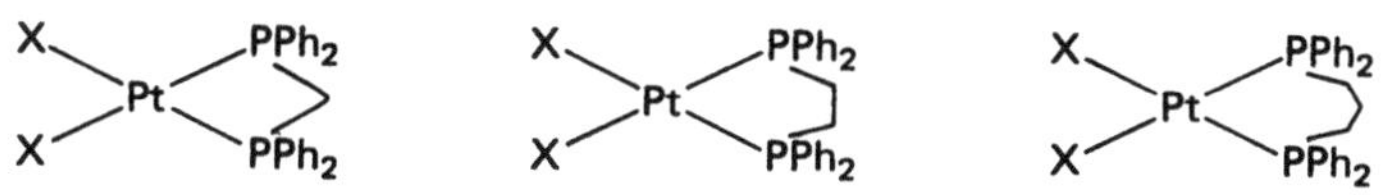

(XXIII) (XXIV) (XXV)

X	(XXIII)	(XXIV) $^1J(Pt,P)$, Hz	(XXV)
Cl	3078	3618	3420
CH_3	1434	1794	1790

e.g. phosphine versus phosphite, affects $^1J(Pt,P)$ in the usual way. The spectroscopic studies concerning this metal cover alkyl and aryl phosphine [*58,134,139-141*] phosphite [*142,143*], phosphonate [*46*] phosphonite [*144*] and other substituted phosphine derivatives [*79,145*]. Worthy of note are a group of papers containing extensive $^2J(P,P)$ data for trimethyl phosphine complexes [*146-148*].

There are many studies which report new classes of platinum phosphine complexes and these often contain interesting, if sometimes camouflaged, NMR data. Since the bias of these investigations is distinctly chemical one often finds a statement of the form: "The value of $^1J(Pt,P)$ is consistent with a structure having two *trans* phosphorus atoms". This type of remark stems from early studies [*58,134*] in which it was found that the unit *trans* "P-Pt-P" often results in a $^1J(Pt,P)$ value of 2200 - 2400 Hz, whereas the *trans* unit "P-Pt-Cl" frequently shows $^1J(Pt,P)$ values in the range 3200 - 3600 Hz. In most cases there is additional evidence (e.g. ^1H NMR and IR data) to support the proposed structure. Work from various groups including ours has produced data that suggest that conclusions based only on the values $^1J(Pt,P)$ should be drawn reluctantly. Consider the complexes *trans*-$PtH(NO_3)(PPh_2Bz)_2$, $^1J(Pt,P)$ = 3190 Hz, *trans*-[PtEt(acetone)$(PPh_2Bz)_2$]BF_4, $^1J(Pt,P)$ = 3302 Hz and *trans*-[Pt($CH_3C{=}CHCH_3$)(CH_3CN)$(PPh_3)_2$]BF_4, $^1J(Pt,P)$ = 3294 Hz. These complexes can be shown to contain *trans* phosphorus atoms; however, the one-bond couplings significantly exceed 3KHz and can fall in the region where one might expect to find a different type of structural unit. Additionally, the presence of other ligands may reduce $^{1}J(Pt,P)$ for a *trans* "P-Pt(II)-P" fragment to a value considerably less than 2200 Hz. Thus in *trans*-$PtCl(SnCl_3)(PEt_3)_2$, $^{1}J(Pt,P)$ = 2042 Hz, the one bond coupling has dropped more than 300 Hz relative to *trans*-$PtCl_2(PEt_3)_2$ ($^{1}J(Pt,P)$ = 2400 Hz). For *trans*-$Pt(SnCl_3)_2(P(OPh)_3)_2$, the one-bond coupling is 3930 Hz, and represents a decrease of almost 500 Hz relative to *trans*-$PtCl_2(P(O(o\text{-}Tol))_3)_2$, $^{1}J(Pt,P)$ = 4405 Hz [*149*].

The complexes XXVI - XXVIII [*150*] provide a nice example which combines the "ring effect" phenomenon with the question of the size of $^{1}J(Pt,P)$ in *trans* phosphine complexes. The phosphorus atoms in

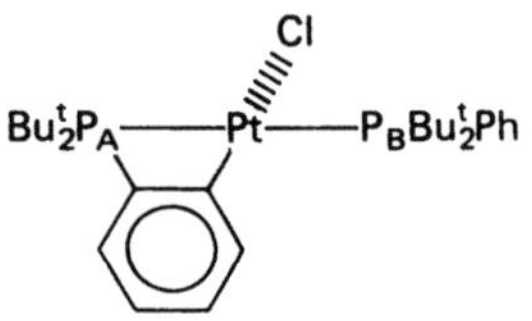

$^{1}J(Pt, P_A)$ = 2222 Hz
$^{1}J(Pt, P_B)$ = 2912 Hz
(XXVI)

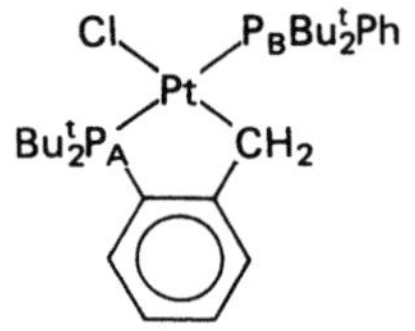

$^{1}J(Pt, P_A)$ = 3262 Hz
$^{1}J(Pt, P_B)$ = 2755 Hz
(XXVII)

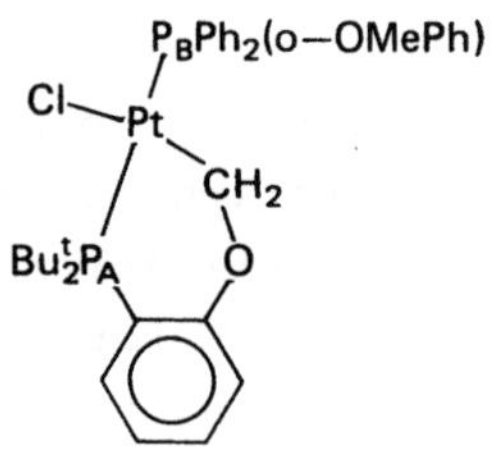

$^{1}J(Pt, P_A)$ = 3178 Hz
$^{1}J(Pt, P_B)$ = 3177 Hz
(XXVIII)

these molecules are "*trans*" and yet the values of the one bond coupling encompass a one kilocycle range. It seems clear that the nature of the remaining coordinated ligands can have a pronounced effect on $^{1}J(Pt,P)$. Although there are no fast rules with regard to structure and this coupling constant, for a given phosphine with identical *cis* ligands, e.g. chloride, a knowledge of the *trans* influence of the fourth ligand can be useful for predicting $^{1}J(Pt,P)$. Conversely, a knowledge of $^{1}J(Pt,P)$ can provide a hint as to what type of ligand might be occupying the fourth position in the square plane when the *cis*-ligands are known, and this latter approach has been used with some success [*151*].

The most common classes of molecule for which $^1J(Pt(II),P)$ values are known include bis [*134*], tris [*45*] and tetrakis [*148*] platinum phosphine complexes of a variety of anionic ligands as well as halogen or oxygen bridged dimers. Under the bis-phosphine heading fall alkyl, [*138*] alkenyl [*152*], acyl [*153*] and hydride derivatives [*154*]. Additionally, there are a number of studies involving polydentate phosphine systems [*131,141,155,156*] as well as sterically large phosphine complexes which readily orthometallate [*157,158*]. The question of ortho-metallation will be taken up in the applications section. A listing of typical $^1J(Pt,P)$ values for a variety of molecular types is shown in Tables 3 - 9 (Chap. G).

Coupling to Main Group Metals

The literature concerned with the values $^1J(M,P)$ for the case where M is not a transition metal is relatively sparse; however, the one-bond metal-phosphorus coupling where the metal comes from group IV a has received some attention. The reduced one-bond phosphorus-carbon [*159*], phosphorus-tin [*36,38*] and phosphorus-lead [*36*] couplings in the molecules PR_3, R = phenyl or alkyl, $Ph_nP(SnMe_3)_{3-n}$ and Ph_2PPbMe_3 are negative. The sign of J for the tin compounds is positive, but inverts in K since the gyromagnetic ratio of tin is negative. Quaternization of the phosphorus atom is generally associated with an *algebraic* increase in $^1K(M,P)$ with $^1K(P,C)$ being noteworthy in that a sign inversion takes place [*159*]. Complexation of the phosphines $Ph_nP(SnMe_3)_{3-n}$ to Cr, Mo and W results in $^1K(P,Sn)$ couplings which approach zero but are still negative [*160*] (e.g. the magnitude of $^1J(Sn,P)$ decreases on complexation). A similar observation concerning the size of $^1J(Sn,P)$ has been made for ligands of the type $PRR'(SnMe_3)$, R,R' = $SnMe_3$, $GeMe_3$, $SiMe_3$, Ph and H, and for the complexes $Ni(CO)_3L$, $Fe(CO)_4L$, $Cr(CO)_5L$, $Mo(CO)_5L$, $W(CO)_5L$, L = $P(SnMe_3)_3$, $PBu^t(SnMe_3)_2$, $PBu^t_2(SnMe_3)$ [*161*]. The negative $^1K(Sn,P)$ and $^1K(Pb,P)$ values are thought to be related to the significantly different energy gap between the phosphorus and metal valence atomic orbitals [*160*]. The values $^1J(Sn,P)$ for the molecules $SnCl_4(PEt_3)_2$ and $SnCl_4(PEt_2Ph)_2$ have been reported and these with other $^1J(Sn,P)$ data are compiled in Table 11 (Chap. G). A study of some one-bond silicon-phosphorus couplings in silylphosphines of the type $Me_xH_{3-x}SiPH_2$, $Me_xH_{3-x}SiPHMe$, $(Me_xH_{3-x}Si)_3P$ has appeared and states that the sign of the reduced coupling in $(SiH_3)_3P$ is negative [*162*].

There have been several reports of coupling between phosphorus and boron in complexes of the type $R_1R_2R_3P\text{-}BH_3$ [*163-166*]. This coupling is believed to be positive and ranges from ∿ 13 to 97 Hz as a function of the substituents on phosphorus. As might be expected substitution of halogen for proton on boron markedly increases this coupling. $^1J(B,P)$ values for Me_3PBX_3, X = H, F, Cl, Br are 64, 174, 166 and 165 Hz, respectively. An empirical correlation relating $^1J(B,P)$ and dative bond strength has been found [*167*]. Several values of $^1J(Al,P)$ for trialkyl phosphine complexes of $AlCl_3$ and $AlBr_3$ have been reported [*168*] as have some values for $^1J(^{17}O,P)$, 90 - 205 Hz [*169*] and $^1J(Se,P)$, 358 - 955 Hz [*170*].

IV Two-Bond Phosphorus-Phosphorus Coupling Constants-Data

Although the interest in $^2J(P,P)$ is somewhat less than that for the one bond interaction, there has been an increase in the number of workers actively engaged in measuring this parameter. In most cases where values of $^2J(P,P)$ are given, the motivation for the measurement stemmed from the expectation that $^2J(P,P)_{trans}$ should be much greater than $^2J(P,P)_{cis}$. In recent work the sign of this coupling has, generally, not been determined, although there is no reason to doubt the conclusion from earlier efforts [*171*] since these have been confirmed by other workers [*172*]. Recently the number and intensity of the ^{13}C signals in bis phosphine complexes has prompted several groups to be concerned with $^2J(P,P)$ and we shall discuss this point further in Chapter 5. Given this renewed interest we have attempted to provide a brief survey of the $^2J(P,P)$ values for a number of different metals and these coupling constants can be found in Tables 8, 18, 19, 20 and 21. Table 18 (Chap. G) has been taken directly from reference 92 which nicely summarizes our knowledge through the early 1970's. The remaining $^2J(P,P)$ data are somewhat more recent and are generally consistent with the theory mentioned earlier.

C Chemical Shifts

A perusal of the ^{31}P NMR literature involving metal complexes reveals that the discussion, offen, centers around the coupling of phosphorus to another nuclei. This statement does not imply a lack of interest in the ^{31}P chemical shift but rather that additional effort has been expended in the direction of coupling constants since these seemed more readily understood. Like most nuclei other than ^{1}H, the normal ^{31}P chemical shift range is large (> 200 ppm) with the screening for a given phosphorus generally thought to be determined by the expressions shown in equation (3.1) [*183*]:

$$\sigma_t = \sigma_P + \sigma_D + \sigma_X \qquad (3.1)$$

in which the total screening, σ_t, consists primarily of contributions from the paramagnetic term, σ_P (which has nothing to do with paramagnetism) and the diamagnetic term, σ_D. Included in the contribution σ_X, are electric field effects, anisotropic effects and the changes in $\delta^{31}P$ resulting from ring currents, solvents, temperature etc.

For nuclei other than ^{1}H, σ_P is recognized to be far and away the most important of the three terms and is generally expressed as is shown in Eq. (3.2) and (3.3). Since we do not routinely have data at our disposal concerning all the singlet to triplet excitations, the equation is often written as shown in (3.4), assuming a mean triplet excitation energy, ΔE. The sum Q_{AB} contains terms representing

$$\sigma^{AA}_{para} = \frac{1}{3}\,(\sigma^{AA}_{xx} + \sigma^{AA}_{yy} + \sigma^{AA}_{zz}) \qquad (3.2)$$

$$\sigma^{AA}_{zz} = -\frac{2e^2h^2}{m^2c^2}\langle r^{-3}\rangle_{3p} \sum_i^{occ} \sum_j^{unocc} \frac{1}{E_i-E_j}(C^A_{i,Px}C^A_{j,Py} - C^A_{i,Py}C^A_{j,Px}) \sum_{B\neq A} q_{AB} \qquad (3.3)$$

$$\text{with } q_{AB} = C^{B}_{i,Px}C^{B}_{j,Py} - C^{B}_{i,Py}C^{B}_{j,Px}$$

$$\sigma_{para} = (\Sigma\Omega_{AB}/\Delta E)<1/r^3> \qquad (3.4)$$

the charge density and bond order between the nuclei A & B, and $\langle r^{-3}\rangle$ is the expectation value of the inverse cube of the non-s orbital radius.

Up to the submission time of this manuscript there appears to have been no clear separation of these three in terms of their relative importance for phosphine complexes. This fact, combined with the paucity of calculations in this area, has led to the elevation of a group of empiricisms to the level of "theory" in the question of the interpretation of phosphorus chemical shifts. This situation will undoubtedly change; however, until such time the empirical approach, often strongly supported by X-ray structure determinations, must suffice. To appreciate how these relationships were developed it is useful to begin with a brief discussion of the phosphine ligands themselves.

Van Wazer and co-workers [*184*] have offered a theoretical interpretation of ^{31}P NMR chemical shifts for phosphine, phosphoryl, thiophosphoryl, selenophosphoryl and phosphonium compounds. Using a treatment which assumes that the ratio of the mean excitation energy to the expectation value of r^{-3} is constant it is concluded that, for the phosphine derivatives PR_3, the C-P-C bond angle is the major parameter affecting the ^{31}P chemical shift. Although this thesis is more than 10 years old a more current theoretical effort supports this idea [*185*]. Indeed, a recent X-ray study concerned with phosphate derivatives has suggested that a direct relation exists between the O-P-O bond angle and the ^{31}P chemical shift [*186*].

The listing below shows clearly that for a few tertiary phosphines as the groups on phosphorus increase in steric size the C-P-C angle opens and the phosphorus chemical shift moves to lower field. Interesting in this context is that when either PPh_3 or PMe_3 are "pinned back" such as in the related phosphine derivatives XXIX, [*187a*] and XXX [*187b*], thus constricting the C-P-C angle a large upfield shift of the ^{31}P resonance is observed [*64*]. Alicyclic trialkyl phosphites do not show the same tendency as the analogous phosphines perhaps due to the flexibility of the P-O-C angles. There are very marked changes in the ^{31}P chemical shifts of the molecules PX_3 as the element X is varied in a given group. Thus for group IV a, X = $C(CH_3)_3$,

PPh_3

−5.6

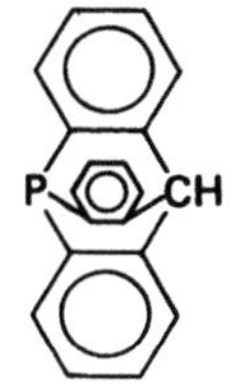

−64.8
(XXIX)

$P(CH_3)_3$

−62

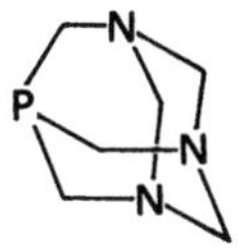

−100
(XXX)

$Si(CH_3)_3$, $Ge(CH_3)_3$ and $Sn(CH_3)_3$, the ^{31}P chemical shift moves to higher field with increasing atomic number of X [*188*]. A formula relating $\delta^{31}P$ to the X-P-X angle and the electronegativity of X has been suggested [*184*]. The effects on the ^{31}P chemical shift of varying the alkyl branches for a wide variety of phosphines have been treated empirically, first by Grim et al. [*189*] and called group contributions, then later by Breen and Quinn [*190*] who express the changes in terms of α, β and γ substituent effects. These terms refer to the changes at phosphorus when a methyl group is substituted at the α, β or γ

$$P_\alpha - C_\beta - C_\gamma - C_\delta$$

positions (see scheme). The effect at phosphorus when a hydrogen is replaced by a methyl group is deshielding (secondary → tertiary phosphine) as is the effect of substituting a methyl for hydrogen at C_β (compare PEt_3 with PMe_3). Of particular interest is the upfield shift observed when a methyl replaces a hydrogen at C_γ . This effect, which is believed to be steric in origin, is in the same sense and of the same order of magnitude (2 - 6 ppm) as that observed in ^{13}C, ^{15}N and ^{75}As [*191,193*] NMR Spectroscopy.

If one now turns to the chemical shift of coordinated trivalent phosphorus for which we find no calculations to date, the most striking empiricism is the linear relation between the coordination chemical shift, Δδ(δ complex - δ free phosphine) and the resonance position of the free phosphine. This correlation has been observed for the complexes $Cr(CO)_5L$, $Mo(CO)_5L$, $W(CO)_5L$, *cis* and *trans* $RuCl_2(CO)_2L_2$, *trans*-$RhCl(CO)L_2$, *mer*-$RhCl_3L_3$, *mer*-$IrCl_3L_3$, $IrCl(CO)L_2$, $IrCl_3(CO)L_2$, $Ni(CO)_2L_2$, *cis* and *trans* -$PdCl_2L_2$ *cis* and *trans* $PtCl_2L_2$ [*194*], where L is a monodentate phosphine ligand. A typical example is shown in Fig. 19. Quite likely as additional data become available other metals will be included.

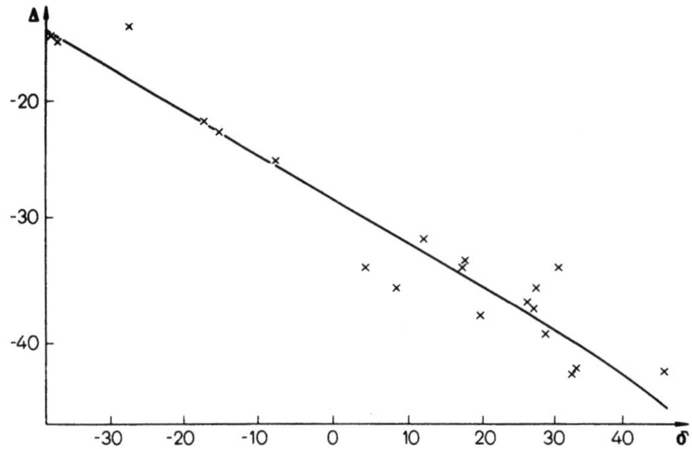

Fig. 19. Plot of the coordination chemical shift, Δ against the ^{31}P chemical shift for the complexes *trans*-$PdCl_2P_2$ [*194*]

There has recently been an attempt to relate the measured C-P-C bond angle to the s-character of the lone pair on phosphorus in the free ligands [*64*]. It is argued that the s-character in the lone pair decreases as the angle opens[9] and that this is associated with a low field shift, e.g. tri-*t*-butyl phosphine has the lowest field ^{31}P-resonance in the following list [*64*].

Compound	Angle	^{31}P,ppm	Cone Angle
PH_3	93.8	- 240	87
PMe_3	98.9	- 62	118
PEt_3		- 20.1	132
PPh_3	103	- 6.0	145
PPr^i_3		20.0	160
PBu^t_3		63.3	182

Since the highest field shifts are associated with ligands with small C-P-C angles the suggestion has been made that the degree to which the C-P-C angle opens on complexation can be related to the

9. This approach may well have some validity; however, it is worth noting the calculations by Lehn and Munsch [*67*] as well as our comments in the previous chapter.

s-character in the phosphorus lone pair in the complex and therefore related to the ^{31}P chemical shift. This is an appealing, if slightly oversimplified picture since there are a large number of cases where other factors contribute significantly to δ^{31}P. As with the free phosphines the complexes often show the methyl substituent effects. Thus from Table 22 we see that, for the complexes $M(CO)_5(PRPh_2)$, M = Cr, Mo, W; R = $CH_{(3-n)}(CH_3)_n$ substitution of a methyl group at C_β produces a downfield shift of 12 - 16 ppm whereas substitution at C_γ in some palladium complexes produces an upfield shift of ~ 2.5-3.0 ppm (Table 24). These effects seem to have some generality as is shown by plots of δ^{31}P in the complexes $Mo(CO_5)P$ against δ^{31}P for a) $Cr(CO)_5P$ b) $W(CO)_5P$ and c) *trans* $PdCl_2P_2$, P = a diphenylalkyl phosphine [*195*]. The rough linearity of these plots, (Fig. 20) suggests that common effects are operating for all the metals. Especially noteworthy is that the slopes of these lines are quite similar. Thus the same type of alkyl substituent effects are functioning for both the complexes and ligands. Unfortunately, there is little known concerning the source of the β methyl substituent effect. Perhaps this is another manifestation of the various intra-ligand angles for different phosphines.

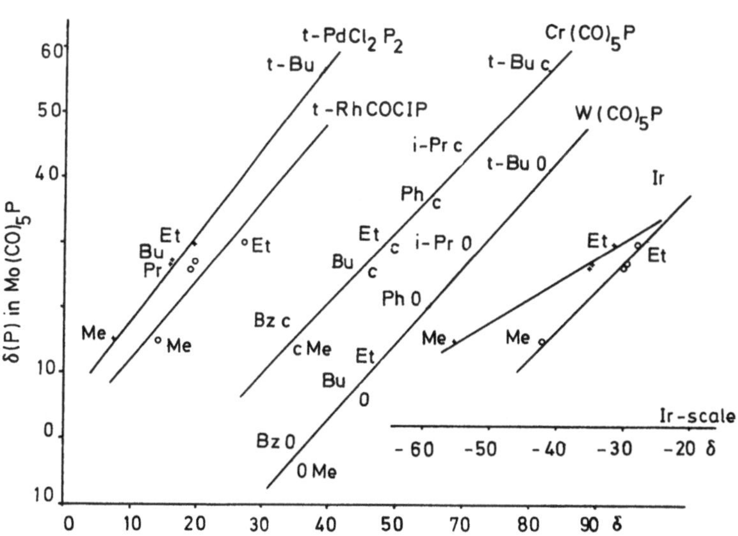

Fig. 20. Plot of δ^{31}P for $Mo(CO)_5P$ against δ^{31}P for a different metal complexes containing the ligand P (P = a diphenylalkyl phosphine). There is a separate high field scale for Ir(III). The upper trace for Ir has P *trans* to Cl, the lower trace P *trans* to P

To this point we have excluded bonding considerations as a source of change for ^{31}P. Not surprisingly the phosphorus chemical shift depends on the nature of the metal and its remaining ligands as well as the nature of the phosphine. In addition, the geometric arrangement of the ligands is important. The effect on $\delta^{31}P$ of changing a single ligand is nicely demonstrated by the complexes *trans*-$PdCl_2$(phosphine)X [*196*] and this is shown in Table 25. For such d^8-square planar complexes one normally thinks of the *trans*- influence (see Chapter 2) of the ligands. As can be seen from the Table an increase in the *trans* influence of the X group shifts the ^{31}P resonance to higher field. It may well be the case that this observation is valid for several other second row transition elements. Thus for $RhCl(PPh_3)_3$ the phosphorus *trans* to chloride has δ = 48.0 ppm whereas the two phosphorus atoms *trans* to one another show δ = 31.5 ppm. [*116*]. Further, it has recently been shown that this *trans* influence criterion may have some validity for five-coordinate ruthenium complexes. [RuH(OH)$(PPh_3)_2$(solvent)] is thought not to have two axial phosphorus ligands since the observed single resonance at 76.2 ppm appears at relatively low field. Phosphorus *trans* to phosphorus in such molecules is thought to be found closer to 40 ppm and phosphorus *trans* to hydride at even higher field [*197*].

For the complexes *trans*-$PtHX(PEt_3)_2$, X = Cl, Br, I, CN, NCO, NCS, SCN, NO_3, NO_2 and N_3, a linear correlation of ν(Pt-H) and ^1J(Pt,P) with $\delta^{31}P$ has been reported [*48*], although in the infra-red series the cyano complex is an exception. Here, as well, it seems that the ability of the group X to coordinate to the metal is also an important factor for the ^{31}P chemical shift. On the other hand the effect on $\delta^{31}P$ of introducing substituents at the para-positions in the complexes *cis* and *trans* $PdCl_2(P(p\text{-}R\text{-}C_6H_4)_3)_2$, R = Cl, H, CH_3, OCH_3, is rather small (< 3 ppm), suggesting that resonance effects through aromatic systems do not markedly affect the phosphorus chemical shift [*198*].

A similar trend is observed for some d^{10} complexes of Ag(I) and Hg(II) (see Table 26). For the complexes HgX_2P_2 P = PPh_3, P(cyclohexyl)$_3$, PBu^n_3, X = NO_3, Cl, Br, I, the phosphorus chemical shift moves to higher field in the order I > Br > Cl > NO_3, which is the order expected based on the coordinating ability (and ionization potential) for the ligand atom [*82*]. In addition as the coordination number of the d^{10} metal increases the ^{31}P value decreases [*51*]. Thus for the complexes $Ag(PTol_3)_2^+$, $Ag(PTol_3)_3^+$ and $Ag(PTol_3)_4^+$ we find ^{31}P chemical shifts (PF_6 salts) of 33.4, 30.6 and 25.6 ppm respectively [*33*], whereas for the mercury complexes $Hg(PMe_3)_2^{2+}$, $Hg(PMe_3)_3^{2+}$ and $Hg(PMe_3)_4^{2+}$ the values are 13.0, 2.5 and - 17.5 ppm, respectively [*51*]. A recent report [*199*] concerned with Pd(O) chemistry was also in line with this idea. Thus $Pd(PPr^i_3)_2$ and $Pd(PPr^i_3)_3$ show values of + 49.3 and + 39.0 ppm respectively, while $Pd(PCy_3)_2$ and $Pd(PCy_3)_3$ are found at + 38.7 and + 25.9 ppm. For $Pd(PEt_3)_3$ and $Pd(PEt_3)_4$ the values are + 9.6 and - 1.5 ppm respectively [*199*] (see Table 27). These are all useful empiricisms but often they must be used with caution. $\delta^{31}P$ for $Hg(CN)_2(PPh_3)_2$ = 17.9 ppm which would suggest, that the cyanide ligand lies between iodide (1.2 ppm) and bromide (22.2 ppm) in its ability to coordinate to mercury and this is certainly not in keeping with other physical data.

A clear sign that ^{31}P chemical shifts are a composite from several sources, comes from a consideration of the effects of geometric isomerism on the phosphorus resonance position. Consider the square planar complexes *cis* and *trans* $-MCl_2(PR_nPh_{(3-n)})_2$ abbreviated as shown

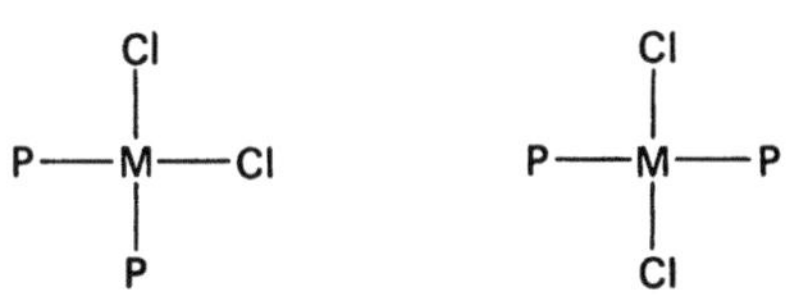

cis (XXXI) *trans* (XXXII)

in XXXI and XXXII. It is well known [*200*] that $\delta^{31}P_{cis}$ appears at lower field by 9-11 ppm than $\delta^{31}P_{trans}$ for M = Pd(II) whereas the reverse is true for M = Pt(II), $\delta^{31}P_{trans} > \delta^{31}P_{cis}$ [*61*]. Since the metal-phosphorus bond distances for these two metals are similar the hypothesis of steric effects as a source of this difference seems inappropriate. Moving to a different coordination number, there is an exceptionally large difference (> 50 ppm) between the chemical shift of axial and equatorial ligands in the five coordinate ruthenium complexes $RuX_2(PPh_3)_3$ with the axial phosphorus at lower field [*201*]. Typical values for these square pyramidal 5-coordinate complexes are 75.7 and 24.1 ppm for X = Cl and 80.0 and 27.4 ppm for X = Br. The trend is reversed on going from the second to the third row in the Periodic Table e.g. the apical phosphorus in $OsCl_2(PPh_3)_3$ has δ = -7.2 ppm whereas the phosphine in the plane shows δ = -1.3 ppm [*201*]. The effect on ^{31}P for the ruthenium complexes is probably the same as that mentioned for $RuH(OH)(PPh_3)_2$(solvent) above, except that now the complex is suggested to have a square pyramidal structure and the low field resonance is assigned to the apical and not the equatorial phosphine.

The oxidation state of a given metal is also of some importance. As mentioned for the Pt(II) complex $PtCl_2(PEt_3)_2$ the ^{31}P resonance of the *trans* isomer lies to low field of the *cis* isomer; however, the reverse is true for the complexes *cis* & *trans* $PtCl_4(PEt_3)_2$. (δ_{cis} = +15.6 ppm, δ_{trans} = +5.6 ppm). The position of the metal in the Periodic Table seems also to affect the resonance frequency of the ^{31}P signal. For the same coordination number, molecular geometry, oxidation state and phosphine, one generally observes a high field shift of the ^{31}P resonance as one descends in a given group. This is illustrated for four different groups of metals in the scheme shown below.

	$\delta^{31}P$	Ref.		$\delta^{31}P$	Ref.
$Cr(CO)_5(PBu^n_3)$	30.2	40	$RuH(CF_3CO_2)CO(PPh_3)_3$	41.0 18.7	202 202
$Mo(CO)_5(PBu^n_3)$	12.2	40			
$W(CO)_5(PBu^n_3)$	- 6.4	40	$OsH(CF_3CO_2)CO(PPh_3)_3$	13.8 0.3	202 202
trans-$PdCl_2(PBu^n_3)_2$	10.0	196	*mer*-$RhCl_3(PBu^n_3)_3$	15.1 - 0.3	31
trans-$PtCl_2(PBu^n_3)_2$	5.1	196	*mer*-$IrCl_3(PBu^n_3)_3$	- 40.3 - 37.7	31

The complexes $M(CO)_4(PPh_2(CH_2)_3PPhR)$, M = Cr, Mo, W show that this effect is not restricted to monodentate systems [*80*]. The difference in ^{31}P chemical shift between the Cr & Mo compounds is approximately 20 ppm as is the difference between the Mo and W complexes. The effect of M is nearly the same for both ends of the bidentate.

An interesting new empiricism, noted by several authors but accented by Garrou [*203*], concerns the "ring contribution" Δ_R to the chemical shift in chelating phosphines. This phenomenon is defined as the difference between the coordination chemical shift, Δ, of a *cis* - substituted bis phosphine complex and the observed Δ value for an equivalent phosphorus in a chelate complex. Put briefly, this effect results in a deshielding of the phosphorus resonance by ∿ 21 - 33 ppm for chelating phosphorus atoms in 5-membered rings whereas this resonances is shifted to higher field by amounts varying from 2 to 25 ppm for four and six membered chelate rings. The classes of compounds under consideration include internally metallated and alkenylphosphine complexes as well as diphosphine chelates and a compilation of data is shown in Table 28. Additional studies have shown that the deshielding effect associated with 5-membered rings can be significantly larger than 30 ppm. Thus for cases where the chelating phosphine atom is involved in two five membered rings, e.g. XXXIII, a deshielding ring contribution, Δ_R, of 80.3 ppm was found [*204*]. For complex XXXIV, where the apical phosphorus takes part in 3 five membered rings,

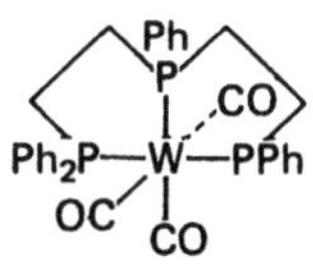

XXXIII

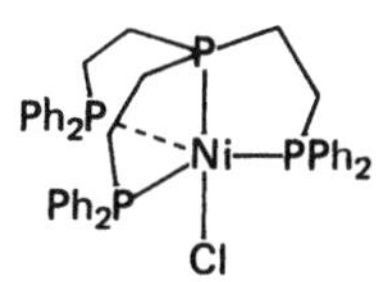

XXXIV

Δ_R = - 157.6 ppm. There is as yet no satisfactory explanation for this phenomenon. Obviously there is still much work to be done before the empiricisms of the previous paragraphs can be viewed in light of equation (3.3). Nevertheless, the available data, taken together, provide a good starting point. A selection of ^{31}P chemical shifts for a variety of metal complexes is given in Tables 22-30.

D Applications

Given the multitude of reports concerned with ^{31}P NMR of metal complexes, it is not difficult to imagine that the number of applications runs into the hundreds. Within our context, it is pertinent to categorize the types of problems which may be addressed using ^{31}P methods as opposed to non-selectively reviewing the literature. We have, therefore, chosen reports covering the different types of chemical problems which one frequently encounters in transition metal chemistry in the hope that the chemical potential of the method emerges clearly at the expense of some detail.

The commonest application of this NMR form takes advantage of the relative spectral simplicity of most $^{31}P\{^{1}H\}$ spectra. The qualitative purity of the reaction product as well as the appearance and disappearance of long-lived intermediates can be readily monitored since each phosphorus type generally affords sharp well resolved resonances. A typical example is given by the equation and has been referred to earlier [*176b*].

trans-$PtH(CH_3CN)(PPh_3)_2^+$ + $MeO_2CC{\equiv}CCO_2Me$ $\xrightarrow{CDCl_3}$

a) *cis*-$Pt(MeO_2CC{=}CHCO_2Me)(PPh_3)_2Cl$[9a]

b) *cis*-$Pt(MeO_2CC{=}CHCO_2Me)(CH_3CN)(PPh_3)_2^+$

c) $Pt(MeO_2CC{\equiv}CCO_2Me)(PPh_3)_2$

The *cis* complexes are readily diagnosed by their AX type spectra. The Pt(O) complex, c, is present only in small quantities and could easily be missed if the ^{31}P spectrum were not readily interpretable. The *cis* complex b eventually isomerizes to *trans* -$[Pt(MeO_2CC{=}CHCO_2Me)(CH_3CN)(PPh_3)_2]^+$ in good yield. An analysis of this reaction based only on the short (or long) term products which are isolated from solution

9a Complex a) derives from the solvent.

would provide considerably less information.

Two further examples in which simple inspection provides relevant chemical information concern the isomers of $Cr(N_2)_2(PMe_3)_4$ [*209*] and $Fe(CO)_4(P(OCH_2)_3P)$ [*92*]. The former could contain either *cis*, XXXVI, or *trans*, XXXV, dinitrogen ligands. The ^{31}P spectrum consists

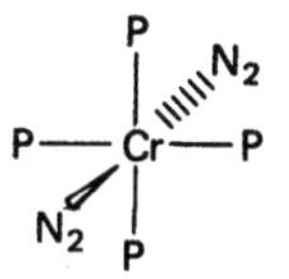
(XXXV)

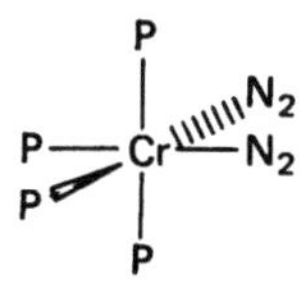
(XXXVI)

of two triplets, $^{2}J(P,P)$ = 16 Hz, thus immediately defining XXXVI as the correct structure. In the iron complex the caged phosphorus derivative may be bound at either (or both) end(s) and as shown in the scheme the large coordination chemical shift readily distinguishes all three possibilities. That end which is bound experiences a considerable deshielding whereas the effect of complexation on the uncoordinated side is much smaller.

	δPO	δPC
$P(OCH_2)_3P$	89.8	- 67.0
ax-$Fe(CO)_4(P(OCH_2)_3P)$	157.4	- 74.4
ax-$Fe(CO)_4(P(CH_2O)_3P)$	87.4	22.4
$Fe(CO)_4(P(OCH_2)_3P)(Fe(CO)_4$	160.4	22.1

Moving to the next level of sophistication, the product of the oxidative addition reaction involving $Pt(PPh_3)_4$ and $SnClPh_3$ has been shown [*87*] to be *cis*-$PtPh(SnClPh_2)(PPh_3)_2$ and not *cis*-$PtCl(SnPh_3)(PPh_3)_2$ as previously reported (e.g. the Sn-C bond is cleaved). Neither of the observed $^{1}J(Pt,P)$ values (2689 Hz for $SnClPh_2$ *trans* to P and 2137 Hz for Ph *trans* to P) was consistent with $^{1}J(Pt,P)$ value for a PPh_3 ligand *trans* to chloride (usually > 3.5 Kc).

The observation of significantly different $^{1}J(Rh,P)$ values (195 Hz and 118 Hz) stemming from the addition of H_2 to $\{RhCl(P(p\text{-}Tol)_3)_2\}_2$ in combination with the ^{31}P line widths in the absence of ^{1}H decoupling led Tolman et al. [*116b*] to conclude that the non-symmetric XXXVII is the reaction product, and not XXXVIII.

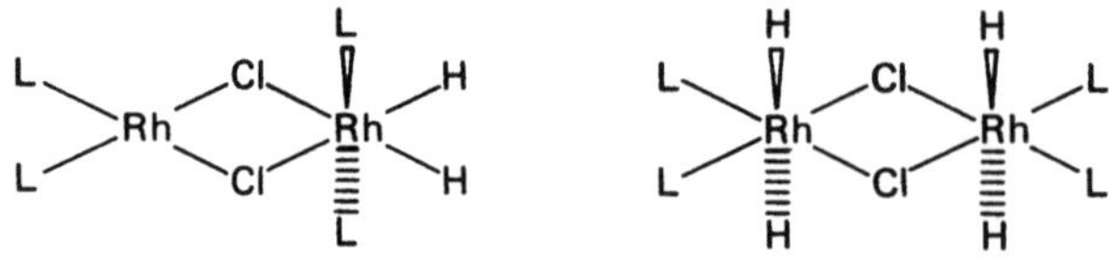

L = P(pTol)$_3$

(XXXVII) (XXXVIII)

The well established empiricisms of ch. 2 point clearly to the presence in XXXVII of both a Rh(I) ($^1J(Rh,P)$ for $(RhCl(P(pTol)_3)_2)_2$ = 196 Hz) and a Rh(III) site. The decision in favor of axial phosphines on the Rh(III) was based on the relatively high field chemical shift of these resonances. These two examples are representative of the classical approach taken by the solution chemist when using $^1J(M,P)$ values.

In addition to the many studies involving the through metal $^2J(X,P)$ couplings, X = ^1H, ^{31}P, there have several recent observations for X = $^{117,119}Sn$, ^{199}Hg and $^{203,205}Tl$ which suggest that this type of coupling may have a wide range of applications. The presence of two distinctly different $^2J(Sn,P)$ values in the compound

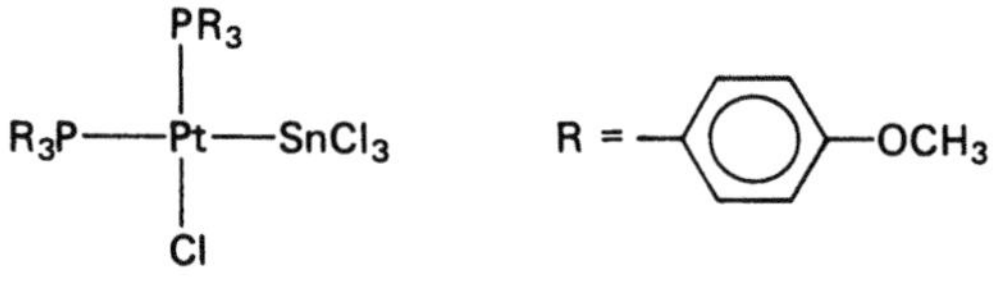

(XXXIX)

$PtCl(SnCl_3)(P(p\text{-}OCH_3C_6H_4)_3)_2$, (4,188 Hz and 207 Hz) is support for the contention that a tin ligand is involved in the coordination sphere of the metal [*201*] and that the structure is as in XXXIX. The coupling of ^{199}Hg to ^{31}P through carbon, in XXXX, (128 Hz) as well

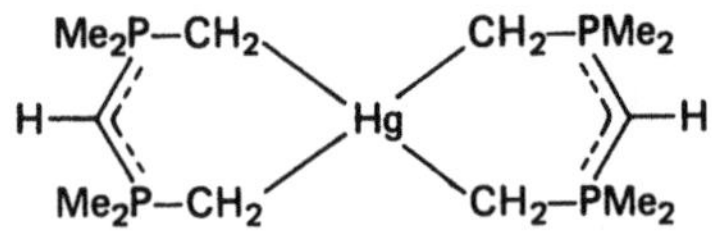

(XXXX)

as a pentet in the ^{199}Hg spectrum helped in the characterization of this molecule as the first tetra-alkyl mercury(II) complex [*211*].

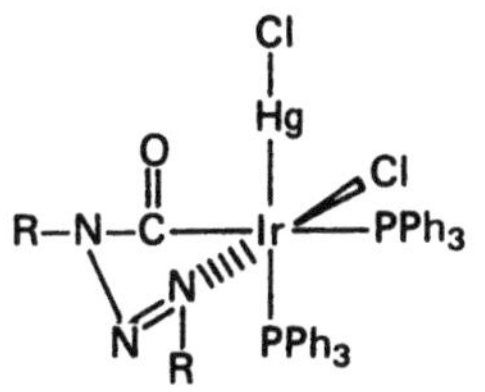

$^2J(Hg, P)_{cis}$ = 397 Hz
$^2J(Hg, P)_{trans}$ = 3 879 Hz
(XXXXI)

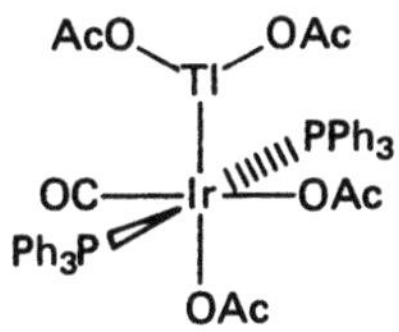

(XXXXII)

Moreover, the identification of the product, XXXXI, resulting from the oxidative addition of a triazenidomercury chloride to *trans* -Ir(CO) Cl(PPh$_3$)$_2$ was made easier by the observation of two markedly differing ^2J(Hg,P) values as well as non-equivalent P atoms (^2J(P,P) = 15 Hz) [*88*]. In a similar vein the observation of ^2J(Tl,P) in XXXXII (1134 Hz) supports the presence of a thallium-iridium bond [*212*], while the calculated value of a three bond coupling J(P,P), ∿ 26 Hz,

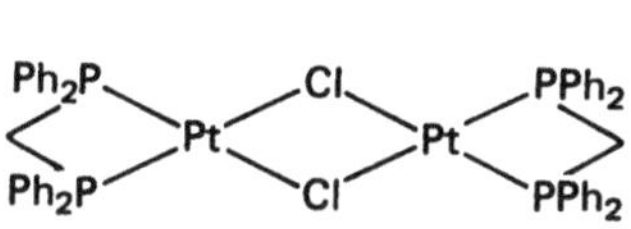

(XXXXIII)

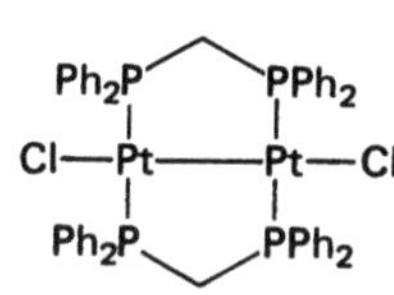

(XXXXIV)

favors the unusual structural type XXXXIV, as opposed to the more conventional type dimer XXXXIII [*213*].

Perhaps as a consequence of the interest in the synthesis of new phosphine complexes for use as homogeneous catalysts, the chemistry of bimetallic phosphine complexes has received a new impetus. These molecules may contain direct metal-metal bonds, or bridging ligands such as in structures XXXXV [*214*], XXXXVI [*128*] or XXXXVII [*110*].

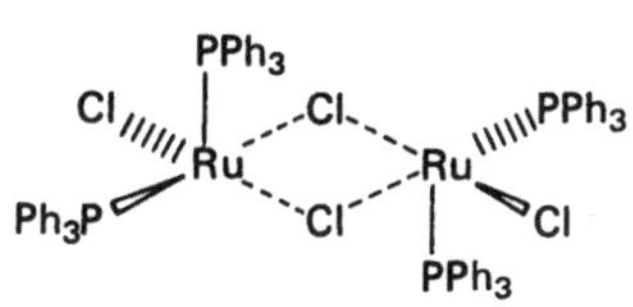

(XXXXV)

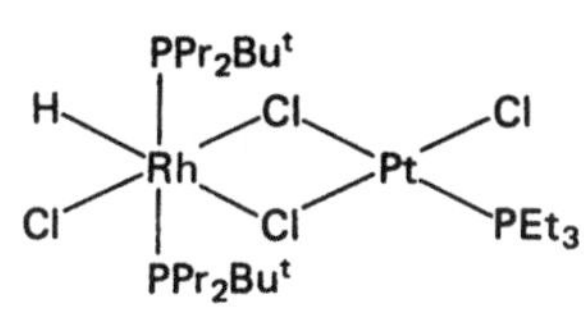

(XXXXVI)

The observation of an AA'BB' (in this case two doublets) ^{31}P spectrum for XXXXV, is consistent with a structure of lower symmetry in which

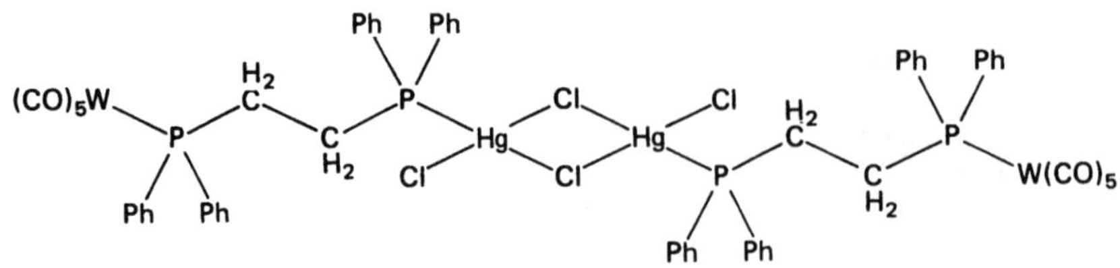

(XXXXVII)

one of the phosphines occupies an axial position. Complexes XXXXVI and XXXXVII both afford ^{31}P spectra containing chemically non equivalent phosphorus atoms each coupled to the appropriate NMR isotope of the metal. Note that in XXXXVII the observed $^{3}J(P\text{-}C\text{-}C\text{-}P)$ value of 47.2 Hz points toward the non equivalence of the ligand atoms within the bidentate while the values $^{1}J(M,P)$ (7,362 and 243 Hz for the mercury and tungsten respectively) indicate that both metals are involved. Undoubtedly, other spectroscopic and chemical methods can be employed to prove the presence of bridging halogen ligands in complexes such as these, but the routine and non-destructive nature of the ^{31}P method would seem to make it a valuable complement, especially where NMR active metals are involved.

The structural chemistry surrounding the metallation of aryl groups, often called "ortho-metallation"[10] since the metal-carbon is formed in a position ortho to a ligating atom, has received increasing attention in the last few years. When a phosphorus ligand is involved changes in the ^{31}P chemical shift and the various phosphorus couplings combine to provide the chemist with a potent structural tool. Heating

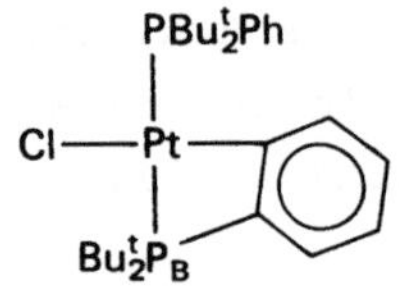

δP_A = 48.6 ppm, $^{1}J(Pt, P)$ = 2912 Hz
δP_B = −31.7 ppm, $^{1}J(Pt, P)$ = 2122 Hz
$^{2}J(P, P)$ = 389 Hz
(XXXXVIII)

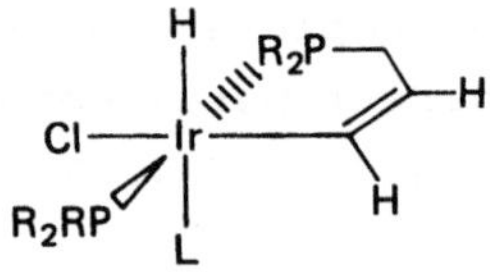

R = Bu^t
L = γ−picoline
$^{2}J(P' P)$ = 365 Hz
(IL)

a solution of *trans* $-PtCl_2(PPhBu^t{}_2)_2$ afforded XXXXVIII (bulky phosphines promote this type of reaction). The $^{2}J(P,P)$ and $^{1}J(Pt,P)$ values

10. The general phenomenon involving the formation of such metal-carbon bonds is commonly called cyclometallation.

suggested *trans* non-equivalent phosphines whereas the unusually high field position of P_B suggests a four membered ring. These results, taken with ^1H and other spectroscopic techniques quickly allow the correct structure to be pieced together [*150*]. An identical approach was used for the product stemming from the reaction of the olefinic phosphine $P(CH_2CH{=}CH_2)Bu^t{}_2$ with $(IrCl(COT)_2)_2$ in the presence of γ-picoline [*215*]. This cyclometallation, forming IL, proceeds with oxidation of the iridium. The value of $^2J(P,P)$ defines the geometric relationship of the phosphorus atoms, while that $^2J(P,H)$ does the same for these atoms with respect to the hydride. The low field position of one of the phosphorus atoms (57.7 ppm) is suggestive of a five membered ring. The same NMR criteria apply for triphenyl phosphite complexes of Fe, Ru, Os, Mn, Co, Pd and Pt [*216*], of which L and LI are examples. In all cases the resonance of the metallated phosphite

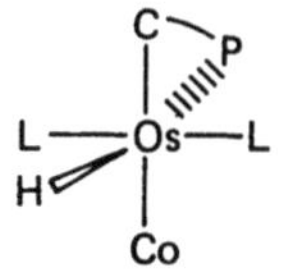

(L)

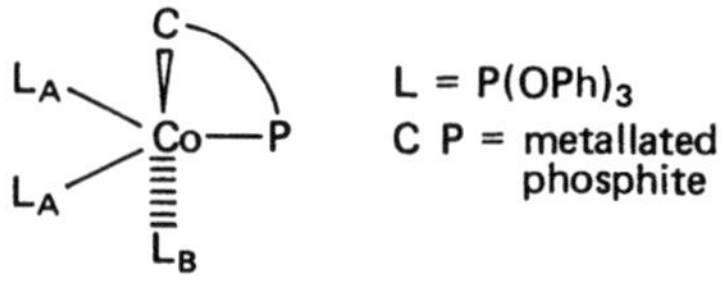

(LI)

appears some 34 to 49 ppm to lower field of the unmetallated phosphite complex as is expected for the P-M-C-C-O five membered ring. In most of these examples ^{31}P NMR is favoured over ^1H studies since the presence of coordinated, but unmetallated phosphite complicates the proton spectrum. Up to this point we have stressed the simplicity of $^{31}P\{^1H\}$ spectra and have more or less restricted our discussion to complexes containing not more than two phosphorus atoms per metal. Not surprisingly the situation is slightly more complicated for complexes stemming from polydentate phosphorus ligands. Consider the ligand QP, LII, and some of its complexes as shown in Fig. 21a. In the octahedral

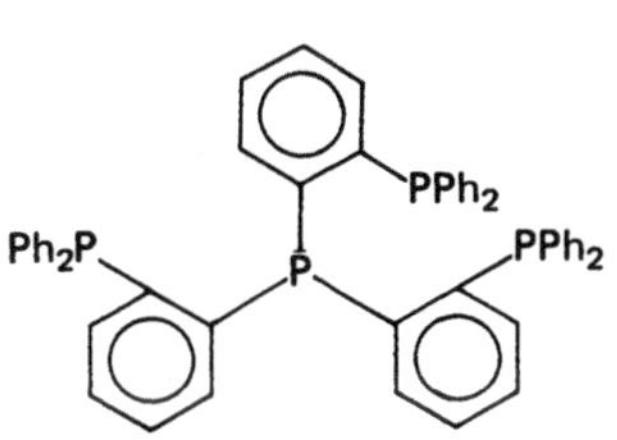

(LII)

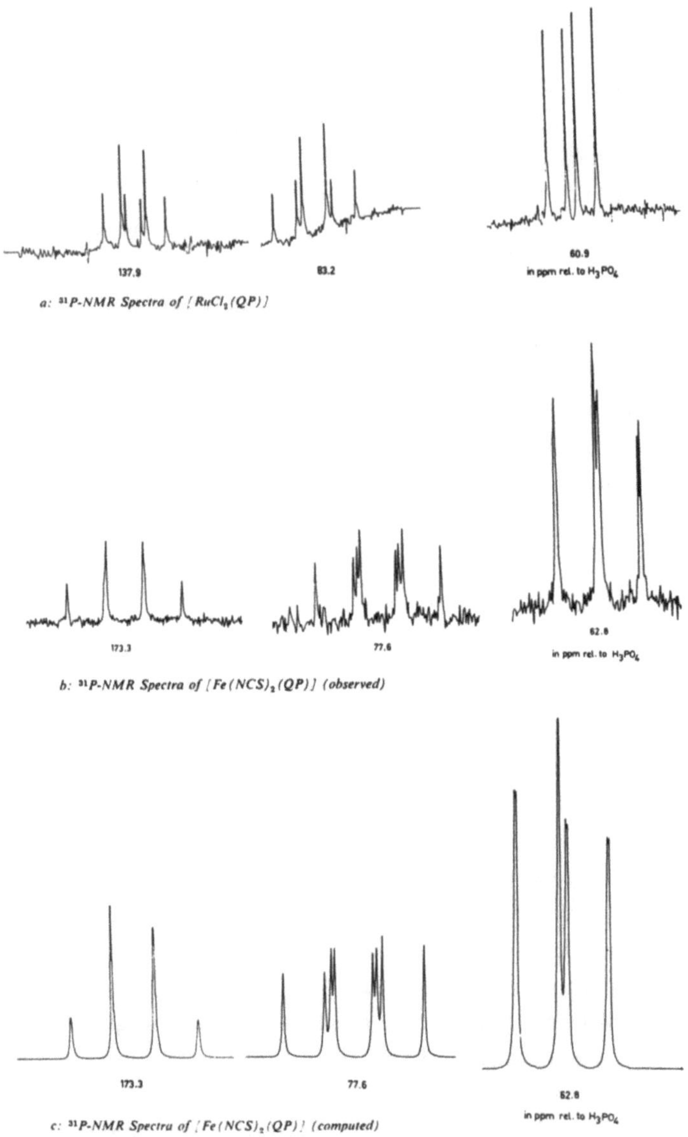

Fig. 21. Spectra of various metal QP complexes. The middle trace deviates from the simple first order expectation [*217*]

$RuCl_2QP$ all four P atoms are coordinated, as in LIII, X = Cl, and the chemical shift differences are sufficient so that the spectrum appears approximately first order. For the $FeCl_2QP$ analog, Fig. 21b, resonances P_B and P_C are 7.5 ppm closer with the result that the spectrum shows second order features [*217*]. The analysis of this type of AB_2X

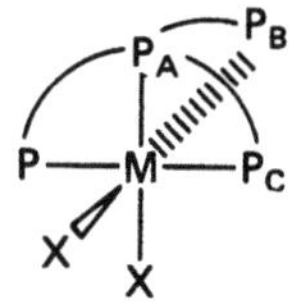

(LIII)

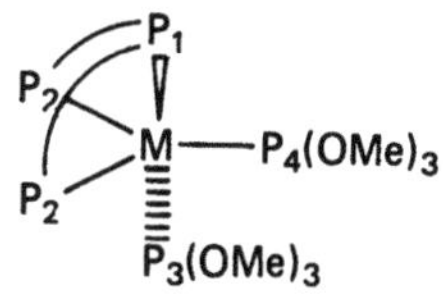

(LIV)

is straightforward and the simulated spectrum is shown in Fig. 21c. For the 5-coordinate fluxional complex Co(etp)(P(OMe)$_3$)$_2$, LIV, where etp is the tridentate bis (2-diphenylphosphinomethyl) phenylphosphine, the spins are more tightly coupled and the ABCD$_2$ spectrum is reminiscent of early ^{1}H studies. The solution of this spectrum reveals that the two P(OMe)$_3$ groups are not being permuted ($^2J(P_1P_3)$ = 193.5 Hz, $^2J(P_3,P_4)$ = 136.5 Hz, $^2J(P_2,P_4)$ = 95.3 Hz) [*218*].

The concept of dynamic NMR and in particular stereochemical nonrigidity has been dealt with previously by many authors. A recent volume [*219*] contains a section relevant to phosphine complexes and is therefore especially noteworthy; however, as we have just introduced a five coordinate potentially fluxional system in which the phosphorus spins are tightly coupled it would be somewhat criminal not to mention the extensive ^{31}P studies stemming from the DuPont group, and directed toward determining the structure and barriers to intramolecular conversion in the ML$_5$ molecules, M = Fe(O), Ru(O), Os(O), [*220a*] Co(I), Rh(I), Ir(I), [*120,142*], Ni(II), Pd(II), Pt(II) [*220b*], L = a trialkyl phosphite. In the slow exchange limit (which often requires temperatures below - 100°C) the ^{31}P spectra are found to be of the type A$_2$B$_3$ (or A$_2$B$_3$X when the metal is involved) which unambigiously defines these molecules as trigonal bipyramidal and not square pyramidal. Moreover, detailed line shape analyses of these second order ^{31}P spectra, assuming a jump model, have led the authors to propose a Berry rearrangement type motion such as depicted in the scheme, in order to explain this intramolecular process. The axial

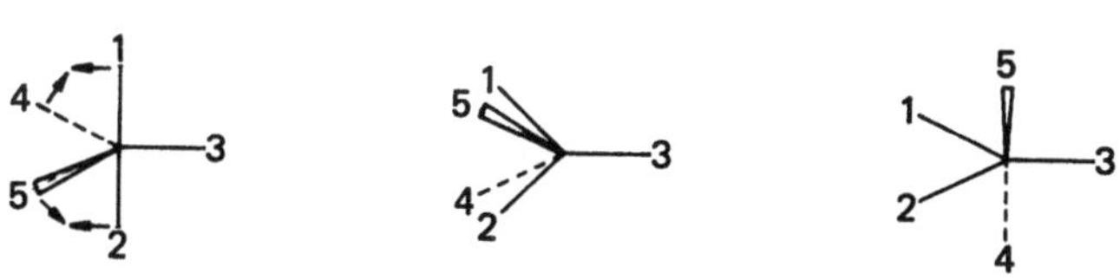

ligands 1 & 2 are exchanged with the equatorial ligands 4 & 5 via a square pyramidal intermediate. Since the possibility of extracting

the mechanistic information depends critically on the observation of the second order phosphorus-phosphorus splittings, the minor annoyance of spectral complexity is actually welcome. A typical example is shown in Fig. 22.

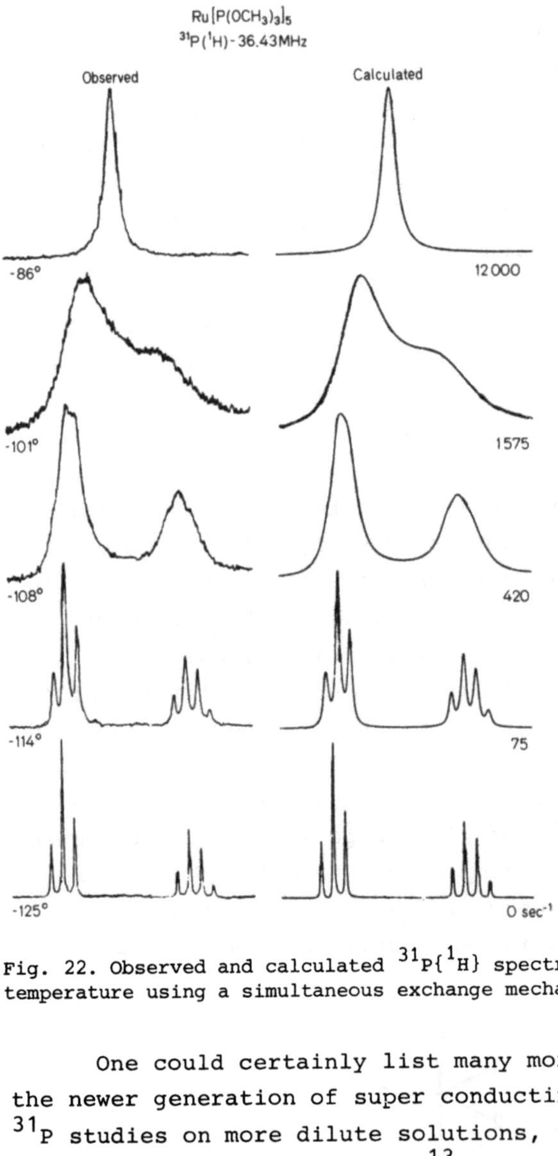

Fig. 22. Observed and calculated $^{31}P\{^1H\}$ spectra for $Ru(P(OMe)_3)_5$ as a function of temperature using a simultaneous exchange mechanism [*220*]

One could certainly list many more applications and given that the newer generation of super conducting magnet systems will allow ^{31}P studies on more dilute solutions, this listing will doubtless blossom further. Just as for ^{13}C, chemists continue to seek simple probes to solve complicated problems and ^{31}P NMR would seem to qualify.

E ^{13}C Studies of Phosphorus Containing Complexes

Although ^{31}P NMR offers the chemist the most direct approach to the study of the phosphorus-metal interaction, information concerning chemical transformations at more remote centers in the molecule can not be monitored as readily. An obvious solution to this problem involves the NMR study of other nuclei within the carbon skeleton of the ligand and of course ^{13}C and ^{1}H NMR techniques are well suited for this purpose. The study of a second nucleus is valuable not only because it may provide a unique method by which a particular problem can be approached, but also because the second study can provide confirmation of a hypothesis stemming from the first study. This is useful since occasionally an NMR study of nucleus B can lead to a completely different conclusion than that which would be drawn from a study of nucleus A alone [*221*]. In view of the enormity of the existing ^{1}H literature we have chosen to limit our few comments to the ^{13}C studies of some phosphorus complexes. Further we have divided these results into two categories: the first concerns the ^{13}C characteristics of the phosphorus ligands themselves and the second the ^{13}C parameters of other ligands in phosphorus complexes.

I Studies of Coordinated Phosphorus-Ligands

One of the earliest ^{13}C studies which reports the chemical shifts and $^{n}J(P,C)$ data for some 18 tertiary phosphines concerns itself with correlating the ^{31}P chemical shifts of these phosphines with the analogous ^{13}C NMR characteristics [*222*]. A plot of $\delta^{13}C$ for the alkanes $R-^{13}CH_3$ against the ^{31}P chemical shift in the molecules $R^{31}PPh_2$ afforded a steady curve. From this and other observations it is concluded that the ^{13}C chemical shifts in these compounds behave similarly to those of related alkanes. The values of $^{1}J(P,C)$ were suggested to provide an estimate the C-P-C bond angle. Although the author estimates an angle for $(t\text{-}Bu)_3P$ of $105^{o}7'$, in agreement with a previous calculation, this result is probably fortuitous since the assumption of a positive value for $^{1}J(P,C)$ in PPh_3 is known to be incorrect [*223*] as is the value given for this same coupling [*224*]. Subsequent work [*224*] has shown $^{1}J(P,C)$ in the free tertiary phosphines to be negative; however, the sign of this coupling is reversed when the phosphorus is quaternized, e.g. (PMe_4^{+} = + 55 Hz) [*223*]. This sign inversion can be relevant for the metal complexes since the phosphorus atoms in these molecules can be thought of as four-coordinate. In support of this suggestion is the observation that $^{1}J(P,C)$ in $Mo(CO)_5(PPh_3)$ has been found to be positive.

The earliest ^{13}C studies on complexes arose from the interest generated by the ^{1}H NMR studies of Shaw and co-workers [*225*]. The assumption made is that *cis* and *trans* phosphine complexes should show single "doublets" and "triplets", respectively, in their ^{13}C spectra, in analogy with what is often observed in the corresponding ^{1}H spectra.

This view promoted an interest in the appearance of the ^{13}C NMR spectra of these complexes. An early statement that ^{13}C NMR would be very useful in the study of the stereochemistry of metal complexes was immediately challenged when "triplets" were observed for some formally *cis* - bis phosphine-π allyl complexes of palladium(II) [*226*]. A clear example showing that generalization can be dangerous is shown in Fig. 23 for $Fe((CH_3)_2PCH_2CH_2P(CH_3)_2)(CO)_3$. Although the two phosphorus atoms are not 180° apart the aliphatic carbon resonances appear as triplets [*227*]. The general question of spectral appearance in palladium(II) complexes was then taken up by Nelson and his co-workers [*228*], some of whose results are shown in Fig. 23. The calculations

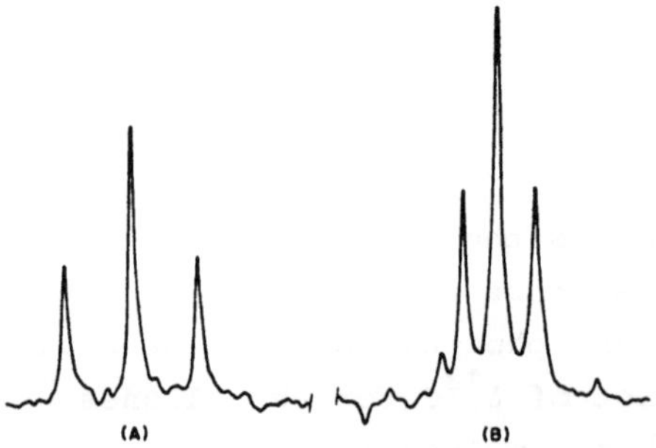

Fig. 23. $^{13}C\{^{1}H\}$ spectrum of $Fe(CO)_3((CH_3)_2P)_2CH_2)$: A) methylene region B) methyl region [*227*]

show that for the A (^{13}C) part of an AXX' system there are a number of possible solutions which lie in between the two extreme cases (doublet and triplet) and indeed several different types have been observed. In our opinion the solution shown in Fig. 24 for J(X,X) = 25 Hz

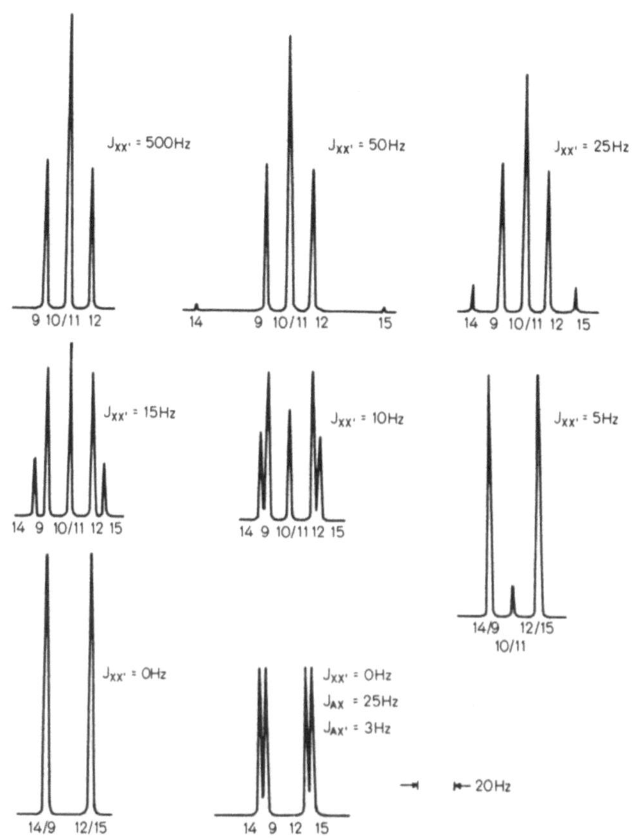

Fig. 24. Some computer simulations for the system $A[X]_2$ where J(A,X) = 25 Hz, J(A,X') = 0.5 Hz and J(X,X') is varied [*228*]

is especially worthy of attention since, unless the spectral S/N is good, this solution could easily be mistaken for a "triplet". From studies [*228*] on these bis phosphine palladium complexes it has suggested that a triplet will be observed in the ^{13}C spectrum of the AAX' system (A = ^{13}C, X,X' = ^{31}P) when:

$$|J(A,X) - J(A,X')|^2 < 8\ J(X,X')\nu\Delta_{1/2} \qquad (4.1)$$

where $\Delta\nu_{1/2}$ is the line width at half height. For the analogous ^{1}H situation, Eq. (4.2) is thought to be valid in order to produce

$$|J(A,X) - J(A,X')|^2 < 2\,J(X,X')\Delta\nu_{1/2} \tag{4.2}$$

the so-called virtually coupled triplet. Our experience [*229*] with this coupling has shown that a) $^{1}J(P,C)$ in complexes of Pd(II) and Pt(II), will often be only 25 - 35 Hz, whereas $^{2}J(P,H)$ may be about one half of these values and b) $\Delta\nu_{1/2}$ ^{13}C may be > 1 Hz. Nevertheless triplet ^{13}C spectra will be more frequently observed if the value J(X,X) falls in an intermediate range. These are useful calculations; however, one should remember that the spin system is actually of the ABX type, since the phosphorus atoms have different chemical shifts when bound to the two isotopes of carbon (see scheme).

$$^{12}C - P_A - \text{Metal} - P_B - {}^{13}C$$

It would be interesting to have some idea of how $^{2}J(P,P)$ varies when the P-M-P angle is greater than 90° but less than 180°. In the section devoted to $^{1}J(M,P)$ both NMR and X-ray data are given for the pseudo tetrahedral complexes $HgX_2(PPh_3)_2$, [*82*] which, unfortunately, are not sufficiently soluble for ^{13}C measurement. However, if one assumes that the analogous PCy_3 complexes show a similar, although not identical, behavior, then the changes in $^{1}J(Hg,P)$ are a qualitative measure of the changing P-Hg-P angle. Despite sizeable changes in the metal-phosphorus coupling (see Table 14) the ^{13}C spectrum of the carbon directly attached to phosphorus is always a triplet [*82*]. It would seem that the ^{13}C spectral appearance should be used with caution when attempting to decide on the complex geometry. Despite the difficulties stemming from the spectral appearance several reports have concerned themselves with the utility of the values $^{n}J(P,C)$ and $\delta^{13}C$ for the determination of complex geometry. Thus for the *cis* and *trans* isomers of $PdCl_2(PR_nPh_{3-n})_2$, the C_α carbon (see scheme) of the *cis* isomer appears at lower field than C_α of the *trans* derivative [*196*]. A correlation relating δC_α to the

$$—\text{Metal} - P - C_\alpha - C_\beta - C_\gamma - C_\delta$$

phosphorus chemical shift in the complexes *trans* -$PdCl_2(PBu_3^n)L$ (L = PBu_3, $PTol_3$, $AsBu_3$, $AsTol_3$, $NHEt_2$, pyridine) has been observed [*196*]. The difference in both the position of C_α and the ^{31}P chemical shift is thought to be related to the *trans*-influence of the group L. There has been a recent report in which the P-C couplings in the linear Pd(0) complexes PdL_2 (L = P(cyclohexyl)$_3$, P(t-Bu)$_2$Ph) have been used to support the suggestion that these complexes are indeed two-coordinate [*230*].

^{13}C NMR has been found to be useful in the question of the stability of the "mixed" complexes $PdCl_2$(phosphine)(amine). Although disproportionation [see Eq.(4.3)] and isomerization can occur in solution, a number of mixed complexes are thought to be relatively stable.

$$2\ PdCl_2(\text{phosphine})(\text{amine}) \longrightarrow PdCl_2(\text{phosphine})_2 + PdCl_2(\text{amine})_2 \qquad (4.3)$$

The ^{13}C NMR spectrum of *trans*-$PdCl_2$(piperidine)($PBu_3{}^n$), shown in Fig. 25, supports this contention as the observation of the three-and

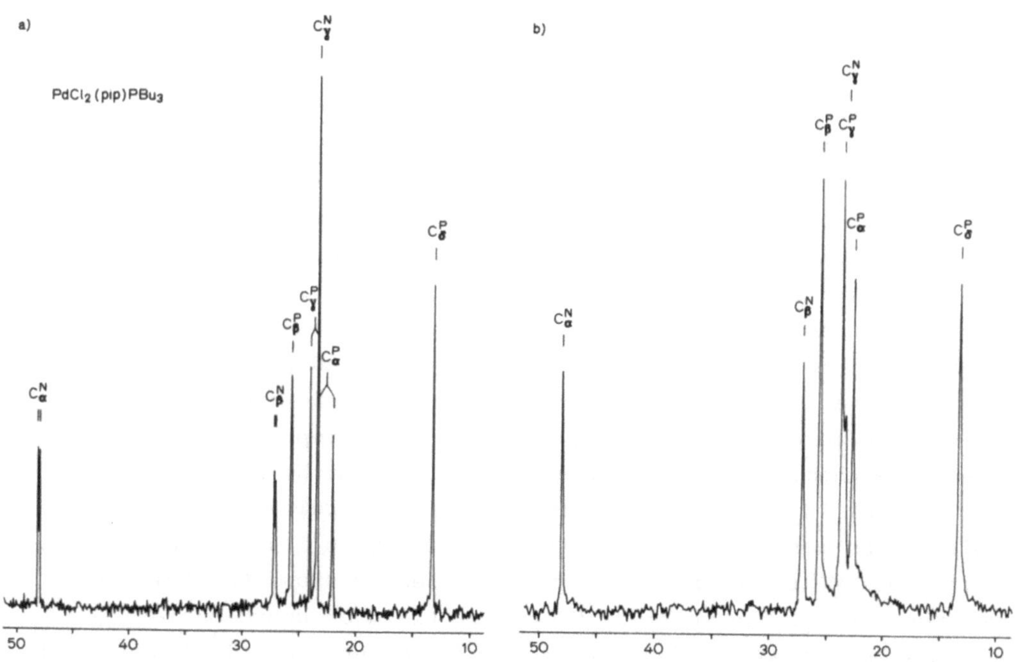

Fig. 25. $^{13}C\{^1H\}$ spectra for *trans*-$PdCl_2$(piperidine)($PBu_3{}^n$) a) without ^{31}P decoupling b) with ^{31}P decoupling [*196*]

four-bond couplings ^3J(*P*-Pd-N-*C*) and ^4J(*P*-Pd-N-C-*C*), respectively, are possible only when the two different types of ligand are bound to the same metal [*229b*]. Before leaving the question of P-C couplings it is worth noting that although $^{1}J(P,C_{\alpha})$ and $^{3}J(P,C_{\gamma})$ are relatively large and easily observable in both the ligands and their complexes, $^{2}J(P,C_{\beta})$ in the complexes is often small or not observed. Since for bis-phosphine complexes, one can only observe $|^{2}J(P,C_{\beta}) + {}^{4}J(P,C_{\beta})|$ it is uncertain whether the two-bond coupling is simply small or if the two-and four-bond couplings have opposite signs. Both points of view have been advanced [*229,231*]. The situation is not definitely resolved although the values $^{2}J(P,C_{\beta})$ in complexes which contain only one phosphine and therefore give simple spectra are less than 4 Hz [*229*].

In rounding off this section, which has been concerned primarily with Pd chemistry, it is worth noting that the methyl chemical shift in the complexes $PdCl_2(P(CH_3)_2(P\text{-}R\text{-}C_6H_4)$ has been shown to correlate with the Hammett σ constant for the group R [*198*] and that studies related to the mechanism of the *cis* to *trans* isomerization in the complexes PdX_2(phosphine)$_2$ have shown ^{13}C NMR to be a useful complement to ^{1}H and ^{31}P NMR [*228*].

One of the most useful applications of ^{13}C NMR spectroscopy concerns itself with the ortho-metallation reaction. This phenomenon, in which a metal-carbon bond is formed, is readily deduced from simple inspection of the $^{13}C\{^{1}H\}$ spectrum. In the unmetallated monosubstituted benzene one finds only four resonances (substituted, ortho, meta and para carbons, respectively), whereas in the metallated ring, which is now an ortho-disubstituted benzene, there is usually no symmetry and all six non-equivalent carbon resonances can be observed. Moreover, the carbon bound to the metal is generally shifted significantly downfield relative to its position in the corresponding uncoordinated complex. The molecules LV [*232a*] and LVI [*232b*] represent typical examples. The chemical shift of the unmetallated ortho carbon in LV is 120.7 ppm. In both of these complexes the coordinated phosphines couple to the metallated carbon thus assisting in its assignment. The spectrum of LVI is shown in Fig. 26 and clearly shows the triplet resonance for C_1. The shift of C_1 to low field on coordination to a metal has been reported for a variety of compounds and is independent of the presence of phosphorus in the complex. The ^{13}C spectra stemming from the product of the reaction of the ylide, $RCOX^{-}Y^{+}$, R = an aromatic group, with metal halides show that ortho-metallation has occurred, presumably after initial coordination of the negative X to the metal [*233*]. The subject of ortho-metallation has been recently reviewed [*234*]. In some cases ortho-metallation is accompanied by

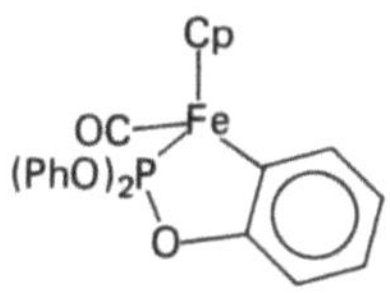

δC_1 = 141.6
J(P, C) = 28 Hz

LV

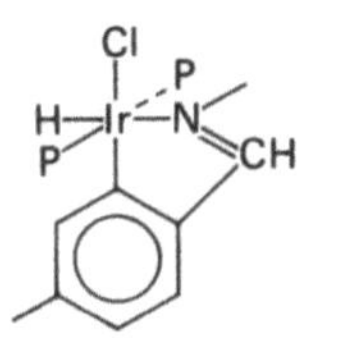

δC_1 = 152.2
J(P, C) = 7.5

LVI

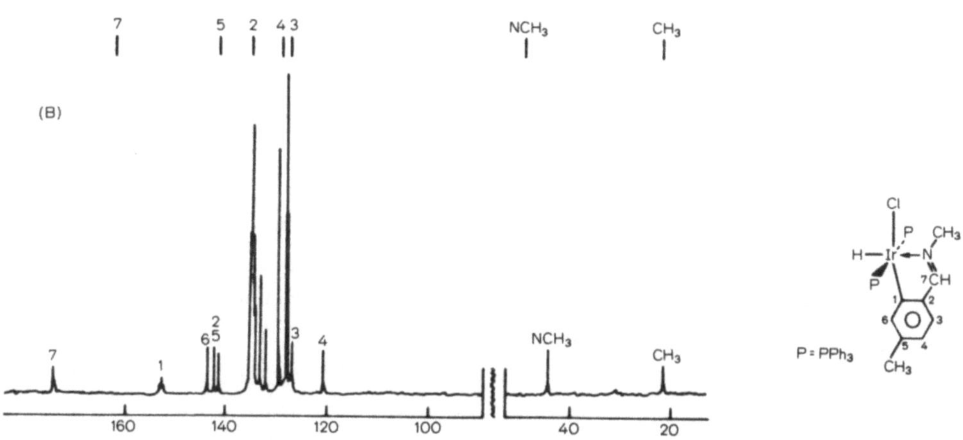

Fig. 26. The ^{13}C spectrum of an orthometallated Schiffs base complex of Ir(III) [*232b*]

characteristic changes in the ^{31}P NMR characteristics of the complexes (e.g. "ring effects" [*203*] on $\delta^{31}P$) and thus both NMR forms are complementary.

In addition to locating the site of metallation, ^{13}C spectroscopy provides a direct probe for locating the site(s) of hydrogen-deuterium exchange. Substitution of a deuterium for a proton generally results in a) loss of ^{13}C signal intensity due to the absence of the proton-carbon induced nuclear Overhauser effect, b) a change in the ^{13}C T_1 value and c) splitting of the ^{13}C resonance by deuterium (I = 1), all of which reduce the signal height. A comparison of the fully protonated spectrum with that of the partially deuterated analog quickly reveals which carbon sites have been affected. In this way the dimers *sym-trans*-$Pt_2Cl_4(PR_3)_2$, R = various alkyl groups, have been shown to incorporate deuterium selectively at the third (C_γ) carbon along the chain [*235*].

II Studies of Non-Phosphorus Ligands in Phosphorus Complexes

The bonding concepts, mentioned in chapter 2 have often been invoked to explain the changes in the ^{13}C spectra of phosphine (and arsine) complexes containing metal-alkyl-carbon bonds; however, the bonding scheme for other ligands such as carbon monoxide, acetylene and olefin complexes is somewhat more complicated in that metal-carbon π-back bonding is generally thought to be important in these systems. For such complexes ^{13}C studies could, in principle, provide a qualitative measure of such interactions.

Thus carbon NMR data have been recently reported for a series of Ni(olefin) $[P(O\text{-}o\text{-}Tol)]_2$ complexes [*236*]. For these Ni(O) complexes it is suggested that increasing back donation from the nickel to the π^* orbital of the olefin can be associated with a larger shift to higher field of the olefin carbon resonance. The appropriate ^{13}C data are shown in Table 31. A correlation between the equilibrium constant K [see Eq.(4.4)] and the coordination chemical shift $\Delta\delta^{13}C$ is also reported [*236*].

$$\text{olefin} + NiL_3 \underset{}{\overset{K}{\rightleftharpoons}} (\text{olefin})NiL_2 + L \qquad (4.4)$$

The ^{13}C spectra for a series of nickel complexes of the type $[L_nNi(CO)_{4-n}]$, when L = PR_3, PPh_2Y. $PPhY_2$, $P(OR')_3$ and PCl_3 have revealed a correlation of $\delta^{13}C{\equiv}O$ with both IR stretching force constants and Kabachinik σ-parameters [*237*]. The carbonyl resonance is found to be deshielded with increasing electron density at the metal (see Table 32). The order of this increase in these complexes is

PCl_3 < $P(OPh)_3$ < $P(OR)_3$ < PPh_3 < PR_3. A similar correlation has been observed for the complexes *cis*-$RuCl_2(CO)_2L_2$, L = a tertiary phosphine [*206*]. The question of platinum to carbon π-back bonding is considered in a recent study by Coulson [*238*]. Based on the carbon NMR parameters of the σ-bond phenyl in the complexes *trans*-$Pt(C_6H_5)X(PEt_3)_2$ the presence of a significant π-interaction is inferred.

Returning to a theme discussed in CH. B, the ^{13}C chemical shifts and the one-bond coupling constants, $^1J(Pt,CO)$, in the platinum(II) carbonyl complexes, $PtX(CO)L_2$, have been found to depend on the *trans*-influence of X [*239*]. When X has a large *trans* influence, couplings in the range 960 - 990 Hz are observed, whereas a smaller *trans* influence produces $^1J(Pt,CO)$ values in the region 1658 - 1817 Hz. Given such an empirical dependence of a ^{13}C NMR parameter, e.g. $^1J(Pt,C)$, one can now estimate the relative *trans* influence of the L groups and the vinyl and acyl groups in the chelate complex LVII.

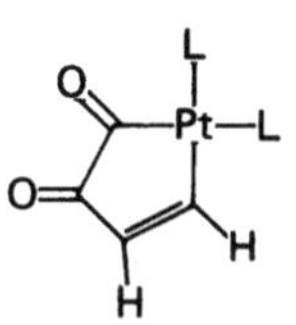

LVII

L = PPh_3, PEt_2Ph, PEt_3, PMe_2Ph, $PMePh_2$, $P(OPh)_3$, (dppe) plus others

Additionally, from the values $^1J(Pt,P)$ the COCO half of the chelate is concluded to have a lower *trans* influence [*130*]. The ^{13}C chemical shifts of some rhenium diphos complexes of type LVIII have been reported. The carbenoid carbon, which has a ^{13}C shift in the region of organic carbonium ions, is especially interesting in that its chemical shift represents a low field extreme for rhenium complexes [*240*].

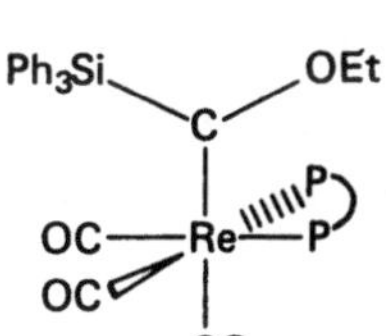

(LVIII)

An extensive group of ^{13}C studies by Clark and co-workers concerned with all of the above themes, e.g. π-bond bonding in olefins, *trans* influence and π-bonding in σ-aryl derivatives has appeared [*241*].

These reports are interesting in that some stand in contradiction to, whereas other are in good agreement with the views expressed in the previous paragraphs. Thus in the complexes $[PtCH_3(COD)L]^+$ the π bonding component of the metal-olefin bond is not thought to be the dominant factor in determining the changes in the ^{13}C resonance positions. In addition, in the phenyl platinum complexes *trans*-$[Pt(C_6H_5)(As(CH_3)_3)_2L]^+$ PF_6^-, σ rather than π interactions are thought to be dominant in the phenyl platinum bond. The question of the NMR *trans* influence is considered in several papers involving methyl platinum complexes [*241*] with the ^{13}C evidence interpreted in a way which supports the σ-rehybridization theory of this effect. Some data for the complexes *trans*-$PtCH_3(PMe_2Ph)_2L^+$ are shown in Table 33.

It is our opinion that all of these reports are probably, at least partially, correct and that the extent to which the various bonding schemes (at least for olefins) contribute will depend markedly upon the metal, its oxidation state and the nature of the remaining ligands within the coordination sphere.

As implied in this section heading our comments on these few ^{13}C NMR studies are only tangentially related to the question of phosphorus NMR. For a more complete picture the reader is recommended to consult several recent reviews [*242*].

Throughout this and the previous chapters we have stressed the sensitivity of the non-proton approach to the transition metal chemistry of phosphorus. Although the emphasis here has been on ^{31}P NMR a very similar type of article might well have been written for ^{13}C and organometallic chemistry. In our view, newer generations of NMR instrumentation will eventually allow us to expand this statement to include other nuclei such as ^{15}N. This advantage is a direct result of our a) monitoring an interaction involving the metal directly and not some probe n-bonds away and b) use of an NMR probe whose chemical shift-structure correlation is more sensitive than that for protons. Considering the vital roles which metals play in both biological and commercial processes, newer methods which help us to probe and measure such interactions should certainly receive due attention.

Appendix A

The Influence of π Orbitals on the Coupling Constant

As mentioned in Chapter 2, there is no direct effect of the π-MO's on the Fermi contact term since this expression includes only s-eigenvectors. Nevertheless there are two mechanisms by which π-orbitals can influence J(A,B):

a) a direct contribution via spin-dipolar and orbital-mechanisms.

b) an indirect mechanism via changing s-eigenvectors in the MO's of molecules of low symmetry for which a clean π-σ-separation is not possible.

a) J(spin-dipolar) and J(orbital)

The orbital and dipolar contributions depend on both, σ- and π-bond orders of non-s-orbitals as shown in formulas (A.1) and (A.2).

$$K_{AB}^{dipolar} = \frac{4\beta^2}{25} \langle r^{-3}\rangle_{P,A} \langle r^{-3}\rangle_{P,B} ({}^3\Delta E)^{-1} x \{ 2 (P_\sigma^2 + P_\pi^2 + P_{\pi'}^2) 3 (P_\sigma P_\pi + P_\sigma P_{\pi'} + P_\pi P_{\pi'}) \} \quad (A.1)$$

$$K_{AB}^{orbit} = \frac{8\beta^2}{3} \langle r^{-3}\rangle_{P,A} \langle r^{-3}\rangle_{P,B} ({}^1\Delta E)^{-1} X \{ P_\sigma P_\pi + P_\sigma P_{\pi'} + P_\pi P_{\pi'} \} \quad (A.2)$$

where β is the Bohr magneton, $\langle r^{-3}\rangle$ are expectation values of valence-p-orbitals, ${}^3\Delta E$ and ${}^1\Delta E$ are triplet and singlet excitation energies, respectively and P_σ and P_π represent the p-orbital σ- and π-bond orders.

Pople and Santry have shown the dipolar contribution to be only a few percent of the total coupling constant involving first row elements (excluding fluorine). Since $\langle r^{-3}\rangle_{np}$ increases as one descends in a group [*243*] (N : 3.02; P : 3.27; As : 7.03; Sb : 9.98; Bi : 18.62) this finding might not be correct for coupling constants between heavy nuclei; however, for the heavier atoms in periods 4 and 5 multiple bonding is less likely than for the atoms of the first two rows. The largest difference is found between the first and second row e.g. it is well known that silicon has less of a tendency to form multiple bonds than does carbon. This can compensate for the increase of $\langle r^{-3}\rangle$ and one can imagine that the direct p-orbital contributions remain below 10% of the total one bond coupling constant involving phosphorus and heavier main group elements. Although the specific case for a bond between phosphorus and a metal is more difficult, results of some recent calculations (see Table A1) involving the phosphorus-tin and phosphorus-lead bonds, for a phosphorus atom in a variety of hybridization states, suggest only a small (< 10%) contribution from these terms. Thus for the chemically significant entries 1-3 in Table A1, it would seem that the spin dipolar contribution is not very important in agreement with a previous suggestion [*21*].

b) π-σ-separation

In the formula for the Fermi-contact term only s-eigenvectors of totaly symmetric MO's are invoked. Consequently one concludes that changes in π-orbitals do not directly influence J. This is only correct for molecules of high symmetry where no π-orbital has the same irreducible representation as the totaly symmetric σ-orbitals. In other words, a clean π-σ-separation is possible.

As one lowers the molecular symmetry, a_1-type π-orbitals appear. These interact directly with a_1-σ-orbitals as shown in Figure A1, thereby changing the s-eigenvectors in the σ-MO's. Thus, a change in the "π"-MO's can influence the "σ"-MO's and therefore the coupling constant. The question of the effect of π-back-bonding on J(M,P) can therefore be determined by the symmetry of the complexes in question.

In Table A2 are shown the symmetry labels of the metal and σ and π ligand group orbitals for some common geometries. Since the metal s-orbital is always of a a_1 or a_{1g} type, only MO's with a_1 or a_{1g} ligand group orbitals contribute to $^1J(M,P)$. Since, for O_h geo-

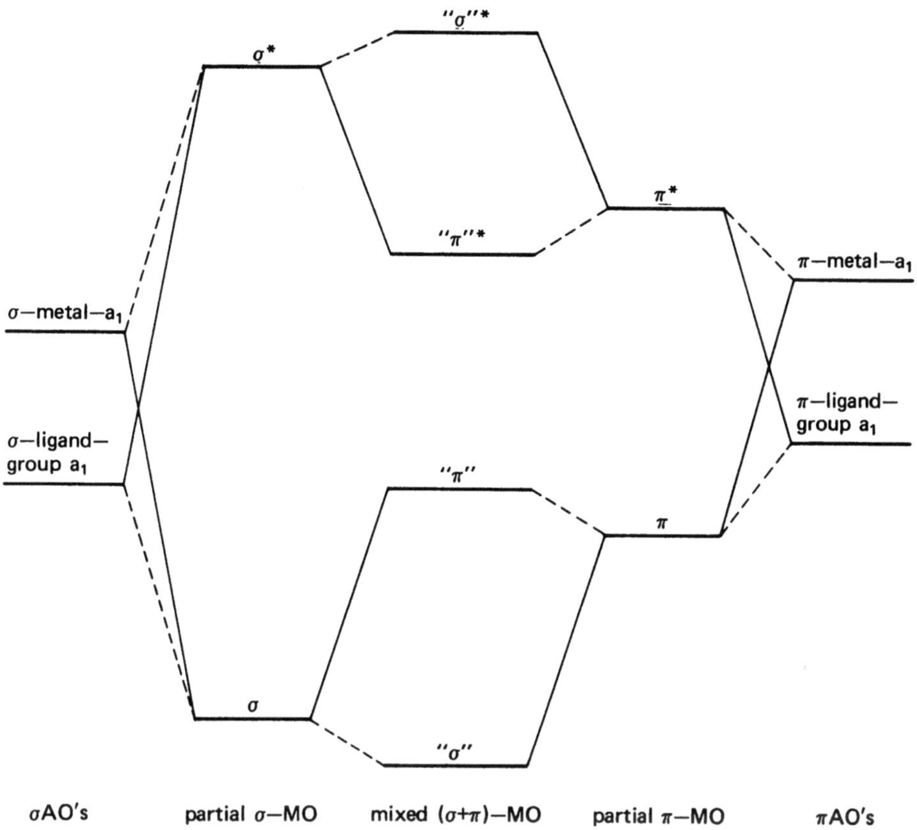

metry there are only σ-ligand orbitals of the a_{1g} type, changes in the π-ligand AO's can only indirectly influence the a_{1g} MO's. This is also true for T_d, D_{4h} and D_3 type geometries; however, for D_{3h} there are two π ligand orbitals of a_1 symmetry. Thus, molecules having this symmetry may have σ and π MO's of the a_1 type which mix, with the result that $^1J(M,P)$ may be directly influenced.

Table A1

Bond	Hybrids used by the Phosphorus	Metal	p-Bondorders p_σ	p_π	$p_{\pi'}$	$K^{dipolar}[10^{20}cm^{-3}]$ Sn(% of 330÷1300)[a)]	Pb(% of 1300)[b)]	$K^{orbital}$ Sn	Pb
P-M	p	p	-1	0	0	13.3(1-4)	25.5(2.0)		
P-M	sp^3	p	$-\sqrt{3}/2$	0	0	20.0(1.5-6.1)	38.4(3.0)		
P-M	sp^3	sp^3	-3/4	0	0	7.4(.6-2.2)	14.4(1.1)		
P=M	sp^2	sp^2	-2/3	1	0	5.9(.5-1.8)	11.4(.9)	-76.8(5.9-23.3)	-147.7(11.4)
P≡M	sp	sp	-1/2	1	1	29.9(2.3-9.1)	57.5(4.4)		

a) $K_{exper.} = 330$ to 1300×10^{20} cm^{-3}.

b) $K_{exper.} = \approx 1300 \times 10^{20}$ cm^{-3}.

Table A2

Geometry	Point-group	s	Symmetry of Metal Orbitals p_z	p_x	p_y	d_{z^2}	$d_{x^2-y^2}$	d_{xy}	d_{yz}	d_{zx}	Symmetry of Ligand Orbitals σ	π
octahedral, ML_6	O_h	a_{1g}		t_{1u} (spanning p_z, p_x, p_y)		e_g (spanning d_{z^2}, $d_{x^2-y^2}$)			t_{2g} (spanning d_{xy}, d_{yz}, d_{zx})		$a_{1g}+e_g+t_{1u}$	$t_{1g}+t_{1u}+t_{2g}+t_{2u}$
tetrahedral, ML_4	T_d	a_1		t_2 (spanning p_z, p_x, p_y)		e (spanning d_{z^2}, $d_{x^2-y^2}$)			t_2 (spanning d_{xy}, d_{yz}, d_{zx})		a_1+t_2	$e+t_1+t_2$
trans-oct., ML_4L_2' / quadratic, ML_4	D_{4h} / D_{4h}	a_{1g}	a_{2u}	e_u (spanning p_x, p_y)		a_{1g}	b_{1g}	b_{2g}	e_g (spanning d_{yz}, d_{zx})		$2a_{1g}+a_{2u}+b_{1g}+e_u$	$a_{2u}+a_{2g}+b_{1u}+b_{1g}+2e_g+2e_u$

oct. with 3 bidentate, $M(L_2)_3$	D_3	a_1	a_2	e		a_1	e		e		$a_{1g}+b_{1g}+e_u$	$a_{2u}+b_{1u}+2e_g+e_u$
trigonal-bipyramidal, ML_5	D_{3h}	a_1'	a_2''	e'		a_1'	e'		e"		a_1+a_2+2e	$2a_1+2a_2+4_e$
cis-oct. ML_3L_3'	C_{3v}										$2a_1+2e$	$2a_1+2a_2+4e$
oct. with centered plane, ML_7	C_{3v}	a_1	a_1	e		a_1	a_1	a_2	e		$2a_1+e$	a_1+a_2+3e
tetrahedral, ML_3L'	C_{3v}										$3a_1+2c$	$2a_1+2a_2+5e$
quadratic pyramidal ML_4L'	C_{4v}	a_1	a_1	e		a_1	b_1	b_2	e		$2a_1+b_1+e$	$a_1+a_2+b_1+b_2+3e$
octahedral ML_5L or *trans* $ML_4L'L''$	C_{4v}										$3a_1+b_1+e$	$a_1+a_2+b_1+b_2+4e$
cis-oct., ML_4L_2'	C_{2v}										$3a_1+a_2+b_1+b_2$	$2a_1+2a_2+4b_1+4b_2$
tetrahedral, ML_2L_2'	C_{2v}	a_1	a_1	b_1	b_2	a_1	a_1	a_2	b_2	b_1	$2a_1+b_1+b_2$	$2a_1+2a_2+2b_1+2b_2$
trig. prism with centered quadratic plane, ML_7	C_{2v}										$3a_1+a_2+2b_1+b_2$	$3a_1+3a_2+4b_1+4b_2$
dodecahedral, ML_8	D_{2d}	a_1	b_2	e_1		a_1	b_1	b_2	e		$2a_1+2b_2+2e$	$2a_1+2a_2+2b_1+2b_2+4e$
quadratic antiprism, ML_8	D_{4d}	a_1	b_2	e_1		a_1	e_2		e_3		$a_1+b_2+e_1+e_2+e_3$	$a_1+a_2+b_1+2e_1+2e_2+2e_3$

Appendix B

Numerical Values of $|\psi_{ns}(0)|^2$

In Chapter two we suggested that the theoretical discussion of J can be partially done on the level of a Pople and Santry formalism. The physical and computational literature is often sufficient such that one should avoid using Eq.(2.5) and resort, wherever possible, to Eq. (2.2). $^3\Delta E$ values for a number of model compounds are readily available from a variety of sources [*244–252*] and changes relative to other related systems can be estimated by the spectrochemical series [*253*].

The literature containing s-density parameters is sometimes overlooked by synthetic chemists although there exist quite complete sets of calculations. Goodman and Raynor have given values for some transition metals [*254*] and Mann and Waber [*255*] published Self-Consistent Relativistic Dirac-Hartree-Fock Calculations for Lanthanide Atoms. A very complete series was published by Froese-Fischer [*256*] for Average-Energy-of-Configuration Hartree-Fock Results for the atoms Helium to Radon. A publication by Mann [*257*] deals with the SCF Hartree-Fock Results for Elements with open shells and for the elements Francium to Nobelium. Although our knowledge about the atomic ground state or near groundstate configuration is quite complete throughout the periodic system, there is, to our knowledge, no systematic study concerning the influences of charge and configuration on the s-densities of the nucleus. Consequently, we have carried out such a systematic study and a summary of these values is given in Table B1. If not otherwise indicated, the values were calculated as given below.

Computional Details

Analytical Hartree-Fock-SCF wavefunctions are used as a basis set which is expressed as a linear combination of Slater-type orbitals.

$$\psi_{i\lambda\alpha} = \sum_p \chi_{p\lambda\alpha} C_{i\lambda p} \tag{B.1}$$

The orbitals are characterized by an index λ, which indicates the symmetry species (λ corresponds to the quantum number 1), by an index α, which indicates the subspecies (the subspecies label the individual members of the degenerate set that transform according to the representation λ) and by index i, which refers to the i-th orbital of symmetry λ. The orbital $\psi_{i\lambda\alpha}$ is expanded in terms of basis functions where the subscript refers to the pth basisfunction of symmetry λ. The basis functions χ are Slater-type orbitals with integer quantum numbers,

$$\chi_{p\lambda\alpha}(r,\theta,\Theta) = R_{\lambda p}(r) Y_{\lambda\alpha}(\theta,\Theta) \tag{B.2}$$

where

$$R_{\lambda p} = [(2n_{\lambda p})!]^{-1/2} (2\zeta_{\lambda p})^{n_{\lambda p}+1/2} r^{n_{\lambda p}-1} e^{-\zeta r} \tag{B.3}$$

and $Y_{2d}(\theta,\Theta)$ are normalized spherical harmonics in complex form. $\psi(r)$ is nonzero for $r\rightarrow 0$ only for s-orbitals and takes the form

$$\psi(0)_{is} = 2C_{is1}(2\zeta_{s1})^{1.5} \frac{1}{2\sqrt{\pi}} \tag{B.4}$$

Comments on Table B1

Only one value of $|\psi(0)|^2$ is given for a given element with a given charge. The s-density depends, within a formal charge, upon the electron configuration and the multiplicity of a configuration. The dependence on the term is small, e.g. for Cr(O) with the configuration d^4s^2, s-densities for the ^3H and the ^1S state are 2.529 and 2.636, respectively. The dependence on the configuration is much more pronounced e.g. for Nb:

Nb^{-1}	d^5s^1	$\lvert\psi(0)\rvert^2$ =	1.650
	d^4s^2		1.788
Nb^0	d^4s^1		3.638
Nb^{+1}	d^3s^1		5.547
	d^2s^2		5.939
	$d^2s^1p^1$		11.740

Thus the promotion of an electron into the next outer shell increases the s-density.

A most interesting point concerns the dependence $|\psi(0)|^2$ on charge. The s-density at the nucleus of a phosphorus 3-s orbital increases 17% for a change of two charge units whereas the corresponding charge increase for rhodium results in a change in the metal 5s-density of 221%. A more detailed analysis of these data is in preparation.

Table B1

Element	$\lvert\psi(0)\rvert^2$ in [a.u.][b)] Charge = -1	0	+1
H		.55	
He		1.439	
Li	.080	.167	
Be		.571	.813
B	1.268	1.411	1.713
C	2.507	2.772	3.205
N	4.482	4.775	5.365
O	7.240	7.637	8.301
F	10.844	11.415	12.250
Ne		16.208	17.257
Na		.529	
Mg		1.214	1.700
Al		2.355	2.745
Si	3.567	3.811	4.288
P	5.312	5.628	6.215
S	7.487	7.916	8.566
Cl	10.124	10.644	11.375
Ar		13.876	14.739

Table B1 (continued)

Element	$\vert\psi(0)\vert^2$ for [a.u.][b)] Charge = -1	0	+1
K	.341	.690	
Ca		1.358	
Sc	1.071	1.408	
Ti	.779[a)]	1.670	
V	.869[a)]	1.950	3.702
Cr	.937	2.216	3.777
Mn	1.096	2.415	4.290
Fe	1.234[a)]	2.573	4.661
Co	1.374[a)]	2.710	5.614
Ni	1.519	2.856	6.159
Cu	1.642	2.961	6.739
Zn		4.541	6.023
Ga	6.621	6.918	7.884
Ge	8.816	9.543	10.608
As	11.697	12.428	13.577
Se	14.903	15.723	16.897
Br	18.419	19.308	20.645
Kr		23.381	24.755
Rb	.667	1.295	
Sr		2.383	3.304
Y	1.713	2.699	4.597
Zr	1.762	3.200	4.827
Nb	1.788	3.638	5.547
Mo	1.990	4.319	6.268
Tc	2.144	4.374	6.966
Ru		4.536	7.446
Rh	2.450	4.650	7.865
Pd	2.598	4.748	8.210
Ag	2.721	4.802	8.603

a) Wavefunctions from *[259]* all other wavefunctions from He to Xe are from *[258]* the values $2C_{is1}$ $(2\zeta_{s1})^{1.5}$ are taken from *[256]* for atoms Cs to Rn and from *[257]* for atoms Fr to Cl.

b) Only one value is given for a given element and a given charge. A complete list of s-densities for other electron configurations is available upon request from R.W.K.

All values are in atomic units [a.u.]. For conversion into e/cm^3 multiply with $6.74834 \cdot 10^{24}$.

Table B1 (continued)

Element	$\|\psi(0)\|^2$ in [a.u.][b)] Charge = -1	0	+1
Cd		6.839	8.900
In	8.899	9.736	11.140
Sn	11.612	12.680	14.130
Sb	14.770	15.773	17.281
Te	18.055	19.197	20.650
J	21.493	22.824	24.364
Xe		26.688	28.283
Cs		1.669	
Ba		2.966	
La		3.092	
Ce		3.211	
Pr		3.329	
Nd		3.446	
Pm		3.563	
Sm		3.680	
Eu		3.799	
Gd		3.918	
Tb		4.039	
Dy		4.161	
Ho		4.284	
Er		4.409	
Tm		4.534	
Yb		4.661	
Lu		5.716	
Hf		6.596	19.392[a)]
Ta		7.393	21.813[a)]
W		8.137	25.748[a)]
Re		8.839	25.834[a)]
Os		9.508	27.660[a)]
Ir		10.151	29.588[a)]
Pt		10.769	31.624[a)]
Au		11.366	33.314[a)]
Hg		11.946	35.074[a)]
Tl		16.280	
Pb		20.531	
Bi		24.862	

Table B1 (continued)

Element	$\lvert\psi(0)\rvert^2$ in [a.u.][b)]		
	Charge = -1	0	+1
Po		29.334	
At		33.980	
Rn		38.821	
Fr		2.544	
Ra		4.396	
Ac		5.322	
Th		6.160	
Pa		5.630	
U		5.776	

Appendix C

Differences in $^1J(M,P)$, $^2J(P,P)_{cis}$ and $^2J(P,P)_{trans}$

In an empirical extention of the LCAO-MO treatment of the one bond coupling constant any n-bond coupling can be related to Eq. (2.2). Only those MO's contribute to the double sum in which both atoms have nonvanishing s-eigenvectors. Thus different parts of a total MO-diagram must be considered in discussing $^1J(P,M)$, $^2J(PP,$ *cis*) and $^2J(PP,$ *trans*). Consider the metal- and ligandgroup orbitals (Fig. 27) of an hypothetical MP_6-molecule:

From these atomic orbitals the MO's of Fig. 28 are derived. For $^1J(MP)$ we require a non vanishing contribution from two orbitals (the MO's must have finite s-character on both the metal and the phosphorus). This is only fulfilled for the a_{1g}-orbitals since in all other MO's there is a nodal plane through the metal and therefore the eigen vector of the metal-s-orbital vanishes. There is only one transition contributing to $^1J(MP)$ as shown in Fig. 28.

For $^2J(PP)$ we require that at both phosphorus positions in question we have nonvanishing σ-ligand group orbitals. In the *cis*-case it is easily seen that this is fulfilled for a_{1g}- and e_g-MO's but not for the t_{1u}-orbitals. This leads to four contributing transitions as shown in Fig. 28. For the *trans* coupling the conditions are also fulfilled for the t_{1u}-orbitals leading to six contributions as shown in the same figure.

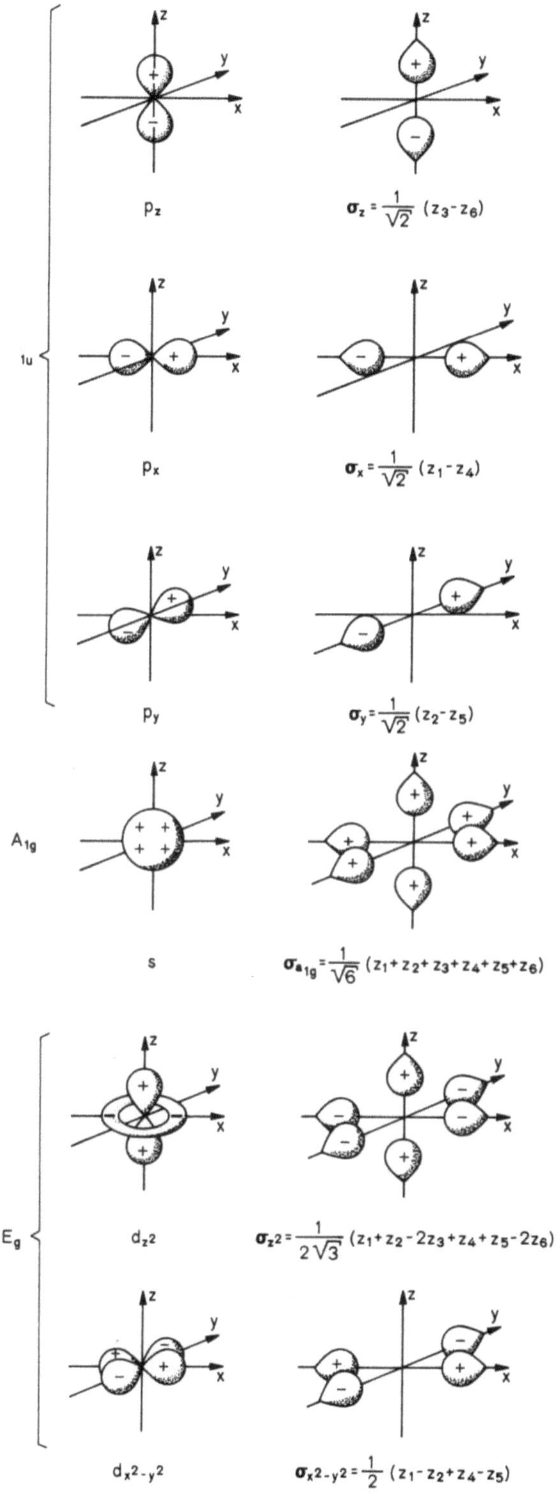

1u
p_z
$\sigma_z = \frac{1}{\sqrt{2}}(z_3 - z_6)$
p_x
$\sigma_x = \frac{1}{\sqrt{2}}(z_1 - z_4)$
p_y
$\sigma_y = \frac{1}{\sqrt{2}}(z_2 - z_5)$
A_{1g}
s
$\sigma_{a_{1g}} = \frac{1}{\sqrt{6}}(z_1 + z_2 + z_3 + z_4 + z_5 + z_6)$
E_g
d_{z^2}
$\sigma_{z^2} = \frac{1}{2\sqrt{3}}(z_1 + z_2 - 2z_3 + z_4 + z_5 - 2z_6)$
$d_{x^2-y^2}$
$\sigma_{x^2-y^2} = \frac{1}{2}(z_1 - z_2 + z_4 - z_5)$

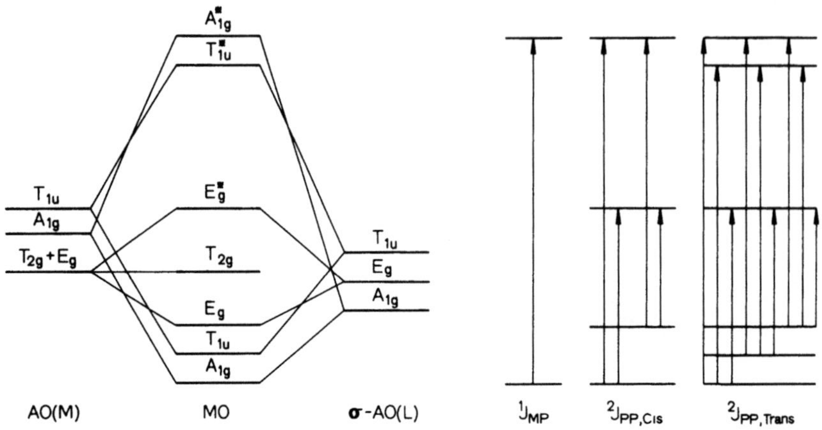
A*1g
T*1u
T1u
A1g
T2g + Eg
E*g
T2g
Eg
T1u
A1g
T1u
Eg
A1g
AO(M)
MO
σ-AO(L)
1JMP
2JPP,Cis
2JPP,Trans

G Tables

Introduction to the Tables

On the following pages are amassed various chemical shifts and coupling constants for transition metal complexes containing phosphorus ligands. The following points are worthy of note:

a) Chemical shifts are in ppm relative to H_3PO_4 with a positive sign indicating a shift to lower field (higher frequency).

b) Coupling constants are in Hz.

c) Solvents and temperatures are not specified and therefore a certain element of "caveat" must reign.

d) The data are only a representative selection. A citation of a given reference is not meant to imply that all of the data from that work is to be found in our compilation.

e) In most, but not all cases, reference is made to the original literature. In some cases a review, or summation of several studies is cited.

Table 1. ^{31}P T_1 and NOE values for some ligands and complexes

Compound	T_1 (sec)	NOE	Compound	T_1 (sec)[b]	NOE[b]	T_1 (sec)[c]	NOE[c]
PEt_3[a]	10.4	0.4	*trans*-$PdCl_2(PEt_3)_2$	14.7	1.16	9.5	0.97
PPr_3^n[b]	15.4	0.35	*trans*-$PdCl_2(PPr_3^n)_2$	9.5	1.18	6.3	1.07
PPh_3	30.8	0.4	*trans*-$PdCl_2(PBu_3^n)_2$	6.0	1.20	4.3	1.14
$PTol_3$[b]	26.7	0.9	*trans*-$PtCl_2(PTol_3)_2$	14.0	1.12	10.4	0.88
$PPh_4^+Br^-$[a]	11.8	0.5	*trans*-$PdCl_2(P(cyclohexyl)_3)_2$	4.6	1.24	3.6	1.21
$P(OPh)_3$[a]	19.8	0.4	*trans*-$PdCl_2(PMe_2Ph)_2$	14.6	1.04		
$PEt_4^+I^-$[a]	9.0	1.0	*cis*-$PdCl_2(PMe_2Ph)_2$	18.5	0.93		

a) Data from Ref. [7a]. Measured in $CDCl_3$ at 29°.

b) Data from Ref. [14]. Measured in CD_2Cl_2 at 30°.

c) Data from Ref. [14]. Measured in $CDCl_3$ at 30°.

Table 2. NOE and T_1 values for selected phosphorus compounds[a)]

Compound	T_1 (sec)	NOE	Ref.	Compound	NOE	Ref.
$PMePh_3Br$	5.95	1.2	15a	PO_4H_3	0.3	15b
PBu^n_4Br	2.3	1.2	15a	$PO_4H_2^-$	0.3	15b
PBu^nPh_3Br	6.5	1.2	15a	PO_4H^-	0.3	15b
$PO(OBu^n)_3$	28.9	1.0	15a	PO_4^{3-}	0.3	15b
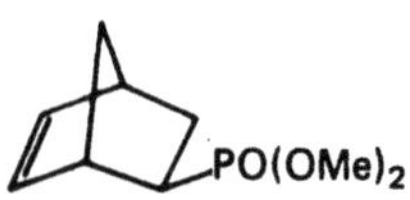	13.2	1.2	15a	HPO_3H_2	0.6	15b
				HPO_3H^-	0.7	15b
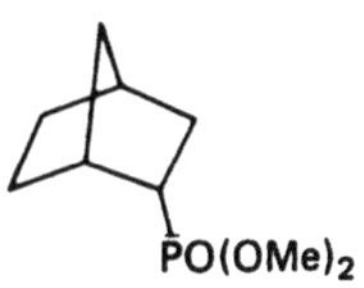	14.8	1.3	15a	HPO_3^{2-}	1.0	15b
				D_2PO_2D	0	15b
				H_2PO_2H	0.4	15b
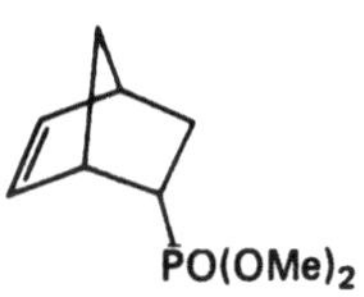	19.0	1.15	15a	$H_2PO_2^-$	1.3	15b
$PO(OMe)_3$	11.6		7c	AMP^{2-}	0.7	15b
$PO(OEt)_3$	18.2		7c	ATP^{4-}	0.1 0.5	15b
$PO(OBu^n)_3$	9.3		7c	$P_2O_7H_4$	0.1	15b
$P(OMe)_3$	6.4		7c	Glycero-phosphate^{2-}	0.6	15b
$P(OEt)_3$	12.4		7c	Glycerophos-phorylcholine	0.7	15b
$P(O\text{-}iPr)_3$	18.5		7c			
$P(OBu^n)_3$	11.8		7c	Phosphoryl-choline	0.3	15b

a) The solvents, temperatures, pH ... vary considerably and the reader should consult the original literature for further details.

Table 3. One-bond platinum-phosphorus couplings in some Pt(O) complexes

Complex	$^1J(Pt,P)$	Ref.
$Pt(PCy_3)_2$	4159	173
$Pt(PMe(t\text{-}Bu)_2)_2$	4015	173
$Pt(PEt_3)_3$	4220	174
$Pt(PTol_3)_3$	4340	174
$Pt(PCy_3)_3$	4236	175
$Pt(PPh_3)_3$	4370	176
$Pt(PEt_3)_4$	3740	174
$Pt(PTol_3)_4$	3780	174
$Pt(C_2H_4)(PPh_3)_2$	3660	177
$Pt(PhCH{=}CHPh)(PPh_3)_2$	3881	178
$Pt(PhC{\equiv}CMe)(PPh_3)_2$	3376 3456	178
$Pt(PhC{\equiv}CH)(PPh_3)_2$	3548 3464	178
$Pt(TolC{\equiv}CH)(PPh_3)_2$	3563 3451	178
$Pt(EtC{\equiv}CEt)(PPh_3)_2$	3430	178
$Pt(PhC{\equiv}CPh)(PPh_3)_2$	3454	178
$Pt(TolC{\equiv}CTol)(PPh_3)_2$	3449	178
$Pt(HC{\equiv}CCH(Me)OH)(PPh_3)_2$	3561 3478	178
$Pt(HC{\equiv}CCH(Ph)OH)(PPh_3)_2$	3547 3479	178
$Pt(PhC{\equiv}CC(Me_2)OH)(PPh_3)_2$	3482 3431	178
$Pt(OHCH_2C{\equiv}CCH_2OH)(PPh_3)_2$	3501	178
$Pt(OHC(Me_2)C{\equiv}CC(Me_2)OH)(PPh_3)_2$	3406	178
$Pt(CF_3C{\equiv}CCF_3)(PPh_3)_2$	3590	136
$Pt[P(OR)_3]_4$		
R		
Et	5386	143
n-Pr	5371	143
i-Pr	5395	143
n-Bu	5388	143

Table 3. (continued)

Complex		$^1J(Pt,P)$	Ref.
$P(OCH_2)_3Et$		5836	143
Ph		5798	143
$Pt[PPh_3][P(OPh)_3]_3$	PPh_3	3430	143
	$P(OPh)_3$	6033	
$Pt[PPh_3]_2$ $[P(OPh_3]_2$	PPh_3	3433	143
	$P(OPh)_3$	6480	
$Pt[\overline{FPOC_6H_4O})_4$		6096	145
$Pt(PF(OPh)_2)_4$		5981	145
Pt(biphos)		3644	79
$Pt(triphos)PPh_3$		3096; 5400	79
$Pt(triphos)PTol_3$		3098; 5380	79
$Pt(triphos)PPh_2Me$		3075; 5370	79
$Pt(triphos)P(OCH_2)_3$-CCH_3		2990; 6787	79
$Pt(triphos)PF_2NMe_2$		2893; 8838	79
$Pt(triphos)P(OPh)_3$		2883; 9150	79
$Pt(triphos)PF_3$		2867	79
Pt(triphos)CO		2837	79

Table 4. One-bond platinum-phosphorus coupling constants as a function of the anionic ligand

X	*trans*-$[PtX_2(Bu_3P)_2]$	$[PtX(Et_3P)_3]ClO_4$		*cis*-$[PtX_2(Bu_3P_2]$
		P *trans* to P	P *trans* to X	
ONO_2	2506	2380	3497	3795
N_3		2312	3124	3409
NCO	2322			3396
NCS	2208			3344
NO_2	2607	2412	2700	2998
CN	2158	2100	2530	
Cl	2400	2233	3499	3504
Br	2327	2255	3490	3477
I	2265	2247	3313	3345

Data from Ref. [*45*].

Table 5. One-bond platinum-phosphorus couplings in a selection of Pt(II) complexes of the type $PtCl_2P_2$

geometry	Phosphorus ligand	$^1J(Pt,P)$	Ref.
cis	PMe_2Ph	3549	61
cis	$PMePh_2$	3616	61
cis	PEt_3	3520	61
trans	PEt_3	2400	61
cis	PEt_2Ph	3530	61
trans	PEt_2Ph	2482	61
cis	$PEtPh_2$	3640	61
cis	PPr_3^n	3530	61
trans	PPr_3^n	2385	61
cis	PPr_2^nPh	3561	61
trans	PPr_2^nPh	2463	61

Table 5. (continued)

Geometry	Phosphorus ligand	$^{1}J(Pt,P)$	Ref.
cis	$PPr^{n}Ph_2$	3639	61
cis	PBu_3^{n}	3500	61
trans	PBu_3^{n}	2392	61
cis	$PBu_2^{n}Ph$	3551	61
trans	$PBu_2^{n}Ph$	2462	61
cis	$PBu^{n}Ph_2$	3641	61
trans	$PBu^{n}Ph_2$	2531	61
cis	$P(OPh)_3$	5793	142a
cis	P(O-*o*-Tol)$_3$	5859	142a
trans	P(O-*o*-Tol)$_3$	4405	142a
cis	P(O-*p*-Tol)$_3$	5840	142a
cis	P(O-*p*ClPh)$_3$	5746	142a

Table 6. $^{1}J(Pt,P)$, Hz, data for the complexes *trans*-$PtHX(PEt_3)_2$

	X	$^{1}J(Pt,P)$	Ref.
1.	Cl	2723	48
2.	Br	2685	48
3.	I	2632	48
4.	CN	2685	48
5.	NCO	2715	48
6.	NCS	2700	48
7.	SCN	2618	48
8.	NO_3	2824	48
9.	NO_2	2768	48
10.	N_3	2730	48
11.	CH_3CN[a)]	2633	176b

Table 6. (continued)

	X	$^1J(Pt,P)$	Ref.
12.	CO[a)]	2443	176a
13.	PPh_3[b)]	2480	179
14.	PEt_3[b)]	2515	179

a) As BF_4^- salts.

b) As ClO_4^- salts.

Table 7. $^1J(Pt,P)$, Hz, values in some *cis* and *trans* complexes of type $PtMeX(PEt_3)_2$

Geometry	X	$^1J(Pt,P)$	
trans	Cl	2821	
trans	Br	2797	
trans	I	2753	
trans	ONO_2	2925	
trans	N_3	2836	
trans	NCO	2800	
trans	NCS	2763	
trans	NO_2	2868	
trans	CN	2595	
		$^1J(Pt,P)$ *trans* to CH_3	$^1J(Pt,P)$ *cis* to CH_3
cis	Cl	1719	4179
cis	Br	1743	4178
cis	N_3	1827	3878
cis	NCS	1686	3994
cis	CH_3	1856	1856

Data from Ref. [47].

Table 8. ^{31}P NMR data[a)] for some metallated phosphorus complexes of the type PtX(P-C)L; (P-C) = metallated, L = ligand

Geometry	X	L	Unmetallated L		Metallated L			Ref.
			δ	$^{1}J(Pt,P)$	δ	$^{1}J(Pt,P)$	$^{2}J(P,P)$	
cis	Cl	$P(OPh)_3$	109.6	3229	100.5	6371	29.0	133
cis	Cl	P(O-*o*-Tol)$_3$	112.7	3201	99.6	6409	30.0	133
cis	Cl	$P(OTol)_3$	112.9	3214	100.4	6407	30.0	133
cis	Br	P(O-*o*-Tol)$_3$	112.0	3200	96.5	6371	30.0	133
trans	Br	PBu^tPh_2	36.9	3098	- 54.2		410	150
trans	Br	PBu^t_2Tol	48.3	2931	- 30.3	2233	389.5	150
trans	Cl	PBu^t_2Ph	48.6	2912	- 31.7	2222	388.5	150
trans	Br	PBu^t_2Ph	49.1	2922	- 36.9	2198	385.5	150
trans	I	PBu^t_2Ph	48.4	2920	- 45.4	2133	380.5	150
trans	NO_3	PBu^t_2Ph	53.4	3070	- 19.5	2312	364.5	150
trans	Br	$PBu^tPr^n_2$	67.4	3086	25.9[b)]	2984[b)]	396.5	150
trans	Cl	PBu^t_3	66.8	2680	- 12.8	2360	382	158

a) Chemical shifts are in ppm (H_3PO_4), coupling constants in Hz.

b) The Pr^n chain is metallated.

Table 9. $^{1}J(Pt,P)$ values for some Pt(IV) complexes

Compound	$^{1}J(Pt,P)$,Hz	Ref.
mer-$[PtCl_3(PMe_3)_3](NO_3)$	+ 2054 *trans* to Cl + 1454 *trans* to P	147
mer-$[PtBr_3(PMe_3)_3](NO_3)$	+ 1975 *trans* to Br + 1461 *trans* to P	147
$(Pr_4N)[PtCl_5(PEt_3)]$	2059	69
$(Pr_4N)[PtCl_5(PBu^n_3)]$	2038	69
cis-$PtCl_4(PEt_3)_2$	2074	69
trans-$PtCl_4(PBu^n{}_3)_2$	2069	69
mer-$(Pr_4N)[PtCl_3(PEt_3)_3]$	2049 *trans* to Cl 1374 *trans* to P	69
trans-$PtCl_4(PEt_3)_2$	1455	69
trans-$PtCl_4(PBu^n{}_3)_2$	1474	69
$PtMe_2Cl_2(PEt_3)_2$	1919	47
$PtMe_2I_2(PEt_3)_2$	1870	47
cis-cis-trans-$[PtCl_2(C_6F_5)_2(PEt_3)_2]$	1463	180
cis-cis-trans-$[PtBr_2(C_6H_5)_2(PEt_3)_2]$	1462	180
cis-cis-trans-$[PtI_2(C_6H_5)_2(PEt_3)_2]$	1803	180
mer-trans-$[PtCl_3(C_6F_5)(PEt_3)_2]$	1475	180
mer-trans-$[PtBr_3(C_6F_5)(PEt_3)_2]$	1481	180
fac-$[PtMe_3I(PMe_2Ph)_2]$	1182	136
fac-$[PtBrMe_3(PMe_2Ph)_2]$	1182	136
fac-$[PtMe_3Cl(PMe_2Ph)_2]$	1180	136
fac-$[PtMe_3I(PMePh_2)_2]$	1080	136
fac-$[PtMe_3I(PMe_3)_2]$	1230	136
fac-$[PtMe_3I(dppe)]$	1100	136
$[PtMe_2I_2(PMe_2Ph)_2]$	1956	136
$[PtMe_2I_2(PEt_3)_2]$	1880	136
$[PtBr_2Me_2(PMe_2Ph)_2]$	1970	136
$[PtMe_2Cl_2(PMe_2Ph)_2]$	2012	136

Table 9. (continued)

Compound	$^{1}J(Pt,P)$, Hz		Ref.
$[PtMe_2Cl_2(PMe_2Ph)_2]$	1190		136
$[PtBr_3Me(PMe_2Ph)_2]$	1710		136
$[PtMe_2(CF_3)I(PMe_2Ph)_2]$	ca. 1200	isomers	136
$[PtMe_2(CF_3)I(PMe_2Ph)_2]$	1086	isomers	136
$[PtBrMe_2(C_3H_5)(PMe_2Ph)_2]$	1160		136
$[PtMe_2(C_3H_5)Cl(PMe_2Ph)_2]$	1230		136
$[PtBrMe_2(CH_2Ph)(PMe_2Ph)_2]$	1300		136

Table 10. One-bond tungsten-phosphorus coupling constants

Compound	$^{1}J(^{183}W,^{31}P)$	Ref.
$W(CO)_5(PPh_3)$	280	40
$W(CO)_5(PMePh_2)$	245	40
$W(CO)_5(PEtPh_2)$	240	40
$W(CO)_5(PPr^{i}Ph_2)$	240	40
$W(CO)_5(PBu^{n}Ph_2)$	250	40
$W(CO)_5(PBu^{t}Ph_2)$	240	40
$W(CO)_5(PBu_2^{n}Ph)$	235	40
$W(CO)_5(PBu_3^{n})$	200, 227	40, 77
$W(CO)_5(PMe_2Ph)$	230	53
$W(CO)_5(P(OMe)_3)$	398	105
$W(CO)_5(P(OEt)_3)$	391	105
$W(CO)_5(P(OPr^{i})_3)$	378	105
$W(CO)_5(P(OBu^{n})_3)$	390	105
$W(CO)_5(P(OPh)_3)$	411, 415.5	77
$W(CO)_5(PPh_2(OMe))$	280	105
$W(CO)_5(P(NMe_2)_3)$	297	105

Table 10. (continued)

Compound	$^{1}J(^{183}W, ^{31}P)$	Ref.
$W(CO)_5(P(NEt_2)_3)$	296	105
$W(CO)_5(P(OCH_2)_3C$-n-Pent)	393	43
$W(CO)_5(P(NMeCH_2)_3C$-n-Pent)	318	43
$W(CO)_5(P(SCH_2)_3C$-n-Pent)	276	43
$W(CO)_5(P(CH_2O)_3C$-n-Pent)	234	43
$W(CO)_5(PF_3)$	485	43
$W(CO)_5(PCl_3)$	426	43
$W(CO)_5(PBr_3)$	398	43
$W(CO)_5(PI_3)$	334	43
$W(CO)_5(P(SnMe_3)_3)$	143	37
$W(CO)_5(P(SnMe_3)_2Ph)$	173	37
$W(CO)_5(P(SnMe_3)Ph_2)$	200	37
$W(CO)_5(\underline{P}Ph_2(CH_2)_2PPh_2)$	240.0	109
$[W(CO)_5(\underline{P}Ph_2(CH_2)_2PPh_2Bu)]PF_6$	243.2	109
$W(CO)_5(\underline{P}Ph_2CH_2PPh_2)$	245.6	109
$[W(CO)_5(\underline{P}Ph_2CH_2PPh_2CH_3)]I$	253.2	109
$W(CO)_5(\underline{P}Ph_2C{\equiv}CPPh_2)$	245.0	109
$W(CO)_5(\underline{P}Ph_2CH{=}CHPPh_2)$	*cis* 238.3 *trans* 239.7 } isomeric olefins	109
$W(CO)_4(PBu^n_3)_2$	*cis* 225, 226.5 *trans* 265, 260.5	77,106
$W(CO)_4(PBu^n_2Ph)_2$	*cis* 220 *trans* 270	106
$W(CO)_4(PBu^nPh_2)_2$	*cis* 230 *trans* 275	106
$W(CO)_4(P(OPh)_3)_2$	*cis* 416 *trans* 446	106
fac-$W(CO)_3(PBu^n_3)_3$	211.0	106
fac-$W(CO)_3(P(OPh)_3)_3$	410.0	106

Table 10. (continued)

Compound		$^1J(^{183}W,^{31}P)$	Ref.
$W(CO)_2(C_5H_5)(SnMe_3)X$			
X			
PMe_3		297.4	108a
PMe_2Ph		302.9	108a
$PMePh_2$		307.8	108a
PPh_3		313.0	108a
$P(OMe)_3$		493.6	108a
$P(OPh)_3$		525.6	108a
$W(CO)_2(C_5H_5)(PbMe_3)X$			
X			
PMe_3		292.3	108b
PMe_2Ph		299.5	108b
$PMePh_2$		305.0	108b
PPh_3		311.5	108b
$P(OMe)_3$		492.4	108b
$W(CO)_2(C_5H_5)CH_3X$			
X			
PMe_3	*cis* *trans*	284.5 236.0	108b
PMe_2Ph	*cis* *trans*	288.0 240.7	108b
PPh_3	*trans*	252.0	108b
$P(OMe)_3$	*cis* *trans*	481.0 460.2	108b
$P(OPh)_3$	*cis* *trans*	520.0 438.5	108b
$W(CO)_4dppm$		202	80d
$W(CO)_4dppe$		231, 239	80d
$W(CO)_4dppp$		222	80d
$W(CO)_4(Ph_2(P(CH_2)_2OPPh_2)$		234 PC 260 PO	80d

Table 10. (continued)

Compound	$^1J(^{183}W, ^{31}P)$		Ref.
$W(CO)_4(Ph_2P(CH_2)_3OPPh_2)$	222 261	PC PO	80d
$W(CO)_4(Ph_2P(CH_2)_2PMePh)$	232 226	PPh PRP	80a
$W(CO)_4(Ph_2P(CH_2)_2PEtPh)$	228 225	PPh PR	80a
$W(CO)_4(Ph_2P(CH_2)_2PPr^iPh)$	229 225	PPh PR	80a
$W(CO)_4(Ph_2P(CH_2)_2PBu^{sec}Ph)$	293 283	PPh PR	80a
$W(CO)_4(Ph_2PCH_2PMe_2)$	199 194	PPh PR	80b
$W(CO)_4(Ph_2PCH_2PMePh)$	200 199	PPh PR	80b
$W(CO)_4(Ph_2PCH_2PPr^iPh)$	200 198	PPh PR	80b
$W(CO)_4(Ph_2PCH_2Pr_2^i)$	203 194	PPh PR	80b
$W(CO)_4(Ph_2P(CH_2)_3PMePh)$	222 216	PPh PR	80c
$W(CO)_4(Ph_2P(CH_2)_3PEtPh)$	222 218	PPh PR	80c
$W(CO)_4(Ph_2P(CH_2)_3PPr^iPh)$	221 220	PPh PR	80c
$WCl(CO)_2(\eta\text{-}C_3H_5)$dppm	168		42
$WBr(CO)_2(\eta\text{-}C_3H_5)$dppm	160		42
$WCl(CO)_2(\eta\text{-}C_3H_5)$dppe	184		42
$WBr(CO)_2(\eta\text{-}C_3H_5)$dppe	187		42
fac-$W(CO)_3$TP	223 224	(apical)	112
fac-$W(CO)_3$QP	228 224 217	(apical)	112

Table 11. One-bond tin-phosphorus coupling constants

Compound	$^1J(^{119}Sn,^{31}P)$, Hz		Ref.
$SnCl_4(PEt_3)_2$	2383.1		38
$SnCl_4(PEt_2Ph)_2$	2255.5		38
$(Me_3Sn)_3P$	+ 832.5	832.3	160, 159
$(Me_3Sn)_3PNi(CO)_3$	424.0		161
$(Me_3Sn)_3PFe(CO)_4$	353.6		161
$(Me_3Sn)_3PCr(CO)_5$	+ 409.5	410.9	160, 161
$(Me_3Sn)_3PMo(CO)_5$	412.3		161
$(Me_3Sn)_3PW(CO)_5$	+ 375.5	376.7	160, 161
$(Me_3Sn)_2PPh$	+ 724	726.4	160, 161
$(Me_3Sn)_2PPhCr(CO)_5$	+ 253		160
$(Me_3Sn)_2PPhW(CO)_5$	+ 216		160
Me_3SnPPh_2	+ 596		160
$Me_3SnPPh_2W(CO)_5$	+ 50		160
$(Me_3Sn)_2PBu^t$	812.3		161
$(Me_3Sn)_2PBu^tFe(CO)_4$	202.0		161
$(Me_3Sn)_2PBu^tCr(CO)_5$	279.8		161
$(Me_3Sn)_2PBu^tMo(CO)_5$	290.1		161
$(Me_3Sn)_2PBu^tW(CO)_5$	242.2		161
$Me_3SnPBu^t_2$	795.4		161
$Me_3SnPBu^t_2Ni(CO)_3$	150.5		161
$Me_3SnPBu^t_2Cr(CO)_5$	128.9		161
$Me_3SnPBu^t_2MO(CO)_5$	145.1		161
$Me_3SnPBu^t_2W(CO)_5$	80.0		161

Table 12. One-bond cadmium-phosphorus coupling constants

Compound	$^{1}J(^{111}Cd,P)$, Hz	Ref.
$CdI_2(PMe_2Ph)_2$	1236.3	34
$CdI_2(PMePh_2)_2$	1123.8	34
$CdI_2(PEt_3)_2$	1308.3	34
$CdI_2(PEt_2Ph)_2$	1238.8	34
$CdI_2(PEtPh_2)_2$	1143.3	34
$Cd_2Cl_4(PBu^t_3)_2$	2168	35
$Cd_2Br_4(PBu^t_3)_2$	2001	35
$Cd_2I_4(PBu^t_3)_2$	1738	35
$Cd_2(SCN)_2(PBu^t_3)_2$	2143	35

Table 13. One-bond mercury – phosphorus coupling constant

Compound	$^{1}J(Hg,P)$, Hz	Ref.
$HgCl_2(PBu_3^n)_2$	5100, 5018, 5018	49, 50, 139
$HgBr_2(PBu_3^n)_2$	4780, 4740, 4951	49, 50, 139
$HgI_2(PBu_3^n)_2$	4040, 4100	49, 50
$Hg(SCN)_2(PBu_3^n)_2$	4500	49
$Hg(CN)_2(PBu_3^n)_2$	4120	49
$Hg(NO_2)_2(PBu_3^n)_2$	5127	139
$Hg(NO_3)_2(PBu_3^n)_2$	5080	139
$Hg(N_3)_2(PBu_3^n)_2$	4978	139
$Hg(NCO)_2(PBu_3^n)_2$	5122	139
$HgCl_2(POc_3)_2$	5160	49
$HgBr_2(POc_3)_2$	4800	49
$HgI_2(POc_3)_2$	3960	49
$HgCl_2(PBu_2^nPh)_2$	5035	50

Table 13. (continued)

Compound	$^{1}J(Hg,P)$, Hz	Ref.
$HgBr_2(PBu_2{}^nPh)_2$	4629	50
$HgI_2(PBu_2{}^nPh)_2$	3726	50
$HgCl_2(PBu^nPh_2)_2$	4754	50
$HgBr_2(PBu^nPh_2)_2$	4216	50
$HgCl_2(PMe_3)_2$	5505	51
$Hg(OAC)_2(PMe_3)_2$	5520	51
$[HgCl(PMe_3)_3]Cl$	3580	51
$[Hg(PMe_3)_4](NO_3)_2$	2255	51
$[CH_3Hg(PMe_3)]Cl$	1875	51
$Hg(NO_3)_2(PPh_3)_2$	5925	82
$Hg(OAc)_2(PPh_3)_2$	5510	82
$HgCl_2(PPh_3)_2$	4675	82
$HgBr_2(PPh_3)_2$	4156	82
$HgI_2(PPh_3)_2$	3074	82
$Hg(SCN)_2(PPh_3)_2$	3725	82
$Hg(CN)_2(PPh_3)_2$	2617	82
$HgCl_2(P(m\text{-}Tol)_3)_2$	4819	82
$HgBr_2(P(m\text{-}Tol)_3)_2$	4292	82
$HgI_2(P(m\text{-}Tol)_3)_2$	3200	82
$Hg(SCN)_2(P(m\text{-}Tol)_3)_2$	3818	82
$Hg(CN)_2(P(m\text{-}Tol)_3)_2$	2752	82
$HgCl_2(P(m\text{-}CF_3Ph)_3)_2$	5303	82
$HgI_2(P(m\text{-}CF_3Ph)_3)_2$	2922	82
$Hg(NO_3)_2(P(m\text{-}CF_3Ph)_3)_2$	6292	82
$Hg(CN)_2(dppe)$	1678	82
$Hg(PO(OEt)_2)_2(dppe)$	576	82
$Hg_2Cl_4(PBu_3{}^n)_2$	7480, 7446	49, 50

Table 13. (continued)

Compound	$^1J(Hg,P)$ Hz	Ref.
$Hg_2Br_4(PBu_3{}^n)_2$	6680, 6624	49, 50
$Hg_2I_4(PBu_3{}^n)_2$	5440, 4348 5120	49, 50
$Hg_2Cl_4(POc_3)_2$	7380	49
$Hg_2Br_4(POc_3)_2$	6620	49
$Hg_2I_4(POc_3)_2$	5380	49
$Hg_2Cl_4(PBu_2{}^nPh)_2$	7514	50
$Hg_2Br_4(PBu_2{}^nPh)_2$	6658	50
$Hg_2I_4(PBu_2{}^nPh)_2$	4250 5020 } mixture	50
$Hg_2Br_4(PBuPh_2)_2$	6553	50
$Hg_2Cl_4(PEt_2Ph)_2$	6627	50
$Hg_2Br_4(PEt_2Ph)_2$	4192	50
$Hg_2I_4(PEt_2Ph)_2$	4985	50
$HgCl_2(P(OEt)_3)_2$	8100	49
$HgI_2(P(OEt)_3)_2$	6420	49
$Hg_2Cl_4(P(OEt)_3)_2$	11,300	49
$Hg_2I_4(P(OEt)_3)_2$	9,060	49
$HgCl(PO(OEt)_2)$	12,830, 12.67 KHz	49, 100
$HgBr(PO(OEt)_2)$	12.24 KHz	49
$HgI(PO(OEt)_2)$	11.18 KHz	49
$HgOAc(PO(OEt)_2)$	12.97 KHz	49
$Hg(PO(OEt)_2)_2$	7,730, 7.50 KHz	49, 100

Table 14. ^{31}P NMR and X-ray data for the complexes $HgX_2(PPh_3)_2$

X	∢ PHgP	∢ XHgX	l(Hg-P)	l(Hg-X)	J(Hg,P)
NO_3^-	131.7	69.6	2.45	2.51	5925
SCN^-	118.1	96.7	2.49	2.57	3725
I^-	109.0	110.4	2.56	2.75	3074
CN^-	108.9	107.8	2.51	2.23	2617
HgJ_2(PSP)	122.8	112.5	2.51	2.81	3420

PSP = $(Ph_2PCH_2CH_2)_2S$.

Data from Ref. [*82*]

Table 15. One-bond silver-phosphorus couplings in some Ag(I) complexes

Compound	$^1J(^{107}Ag,^{31}P)$, Hz[a)]	Ref.
$[Ag(P(p\text{-}Tol)_3)_4]X$		
<u>X</u>		
F	224	33
Cl	225	33
Br	230	33
B_3H_8	230	33
S_2PF_2	225	33
CF_3COO	230	33
NO_3	224	33
PF_6	225	33
ClO_4	225	33
$[Ag(P(OEt)_3)_4]X$		
<u>X</u>		
Cl	341	33
Br	340	33

Table 15. (continued)

Compound	$^1J(^{107}Ag, ^{31}P)$, Hz[a)]	Ref.
I	340	33
SCN	342	33
NO_3	342	33
ClO_4	341	33
$[AgX(P(p\text{-}Tol)_3)_3]$		
X		
F	273	33
Cl	280	33
Br	278	33
I	266	33
CN	230	33
SCN	∼270	33
B_3H_8	300	33
S_2PF_2	308	33
CF_3COO	310	33
NO_3	312	33
PF_6	321	33
$B_9H_{12}S$	316	33
$[AgX(P(OEt)_3)_3]$		
X		
Cl	408	33
Br	398	33
I	390	33
CN	324	33
SCN	397	33
NO_3	450	33
ClO_4	472	33

Table 15. (continued)

Compound	$^{1}J(^{107}Ag, ^{31}P)$, Hz[a)]	Ref.
$[AgX(P(p\text{-}Tol)_3)_2]$		
<u>X</u>		
F	450	33
Cl	378	33
CN	278	33
B_3H_8	390	33
S_2PF_2	410	33
CF_3COO	451	33
NO_3	470	33
PF_6	496	33
ClO_4	503	33
$B_9H_{12}S$	420	33
$[AgX(P(OEt)_3)_2]$		
<u>X</u>		
Cl	550	33
Br	536	33
NO_3	680	33
ClO_4	756	33
$Ag(PBu_3^n)BF_4$	759	95
$Ag(PBu_3^n)_2BF_4$	470	95
$Ag(PBu_3^n)_3BF_4$	304	95
$Ag(PBu_3^n)_4BF_4$	219	95
$Ag(NiS_2C_2(CN)_2)(PEt_3)_2$	426	98
$Ag(NiS_2C_2(CN)_2)(PEt_2Ph)_2$	426	98
$Ag(NiS_2C_2(CN)_2)(PEtPh_2)_2$	413	98
$Ag(NiS_2C_2(CN)_2)(PPh_3)_2$	394	98

Table 15. (continued)

Compound	$^{1}J(^{107}Ag,^{31}P)$, Hz[a)]	Ref.
AgX(PP)		
X		
I	378	97
Cl	411	97
NO_3	461	97
BF_4	515	97

a) Values are for $^{1}J(^{107}Ag,^{31}P)$ only. The $^{1}J(^{109}Ag,^{31}P)$ values approximately satisfy the relation $^{1}J(^{109}Ag,^{31}P) = {}^{1}J(^{107}Ag,^{31}P)(\gamma_{^{109}Ag}/\gamma_{^{107}Ag})$.

Table 16. One-bond rhodium-phosphorus coupling constants

Compound	^{1}J(Rh,P), Hz		Ref.
	P *trans* to anion	P *trans* to P	
$RhCl(PPh_3)_3$	189	142	116a
$RhBr(PPh_3)_3$	192	141	116a
$RhI(PPh_3)_3$	194	139	116a
$RhCl(PCy_3)_2$(S=CNPh)	104.4		32
$RhCl(PCy_3)_2$(S=CNTol)	104.6		32
$RhCl(PCy_3)_2$(S=CO)	104.9		32
$RhCl(PCy_3)_2$(O=O)	106.3		32
$RhCl(PCy_3)_2$(S=CNMe)	106.5		32
$RhCl(PCy_3)_2$(S=CS)	106.6		32
$RhCl(PCy_3)_2$(S=CN-Allyl)	106.6		32
$RhCl(PCy_3)_2$(PhN=CO)	110.5		32
$RhCl(PCy_3)_2$(TolN=CNTol)	110.6		32
$RhCl(PCy_3)_2$(PhC≡CPh)	115.4		32
$RhCl(PCy_3)_2$(HC≡CH)	116.0		32
$RhCl(PCy_3)_2(H_2C{\equiv}CH_2)$	118.6		32

Table 16. (continued)

Compound	$^{1}J(Rh,P)$, Hz P *trans* to anion	$^{1}J(Rh,P)$, Hz P *trans* to	Ref.
$RhCl(PCy_3)_2(EtC{\equiv}CEt)$	121.3		32
$RhCl(PCy_3)_2(SO_2)$	107.6		32
$RhCl(PCy_3)_2(CO)$	119.9		32
$RhCl(PCy_3)_2(N_2)$	122.0		32
$RhCl(PCy_3)_2(CS)$	132.1		32
$RhCl(PCy_3)_2(PhNO)$	153.1		32
$[RhCl(P(p\text{-}Tol)_3)_2]_2$	195		32
$RhCl(P(p\text{-}Tol)_3)_3$	189	143	116b
$RhCl(C_2H_4)(P(p\text{-}Tol)_3)_2$		129	116b
$RhCl(C_2H_4)(PPh_3)_2$		128	116b
$RhCl(CO)(PPh_3)_2$		124	116b
$RhCl(CO)(P(p\text{-}Tol)_3)_2$		124	116b
$Rh(O_2CCF_3)(CO)(PPh_3)_2$		137	117b
$Rh(OAc)(CO)(PPh_3)_2$		132	117b
$RhCl(CO)P_2$			
<u>P</u>			
PMe_2Ph		117.7	29
$PBu^n{}_3$		115.9	29
PMe_2Bu^t		117.6	29
$PMePh_2$		122.4	29
$PPr^n{}_2Ph$		119.6	29
$PBu^n{}_2Ph$		120.0	29
PEt_3		116.1	29
PEt_2Ph		120.4	29
$PEtPh_2$		123.3	29
$PPr^n{}_2Bu^t$		118.9	29

Table 16. (continued)

Compound	^{1}J(Rh,P), Hz		Ref.
	P *trans* to anion	P *trans* to	
PPh_3		124	29
$PBu^n{}_2Bu^t$		119.1	29
PEt_2Bu^t		119.3	29
$PPr^nBu^t{}_2$		120 $\pm$ 3	29
P(p-Tol)$Bu^t{}_2$		126.4	29
$PPhBu^t{}_2$		125.3	29
$P(OMe)_3$		193.9	29
RhCl(CO)$(PBzPh_2)_2$		118	181
RhCl(CO)(PP)		124	181
RhBr(CO)(PP)		123	181
RhI(CO)(PP)		122	181
Rh(NCS)(CO)(PP)		125	181
Rh(CH_3CN)(CO)(PP)		120	181
RhCl(CO)$(PClPh_2)_2$		122	117c
RhCl(CO)(dppe)	158 124		117c
RhBr(CO)(dppe)	160 125		117c
[Rh$(dppe)_2$]Cl		133	117c
fac-$RhCl_3(PMe_3)_3$	113		30
fac-$RhCl_3(PEt_3)_3$	114		30
fac-$RhCl_3(PBu^n{}_2Ph)_3$	122		30
fac-$RhCl_3(PPr^n{}_2Ph)_3$	117		30
mer-$RhCl_3P_3$			
PEt_3	112.1	83.8	31
$PPr^n{}_3$	111.9	83.8	31
$PBu^n{}_3$	112.5	83.6	31
PMe_2Ph	112.3	84.6	31

Table 16. (continued)

Compound	$^{1}J(Rh,P)$, Hz		Ref.
	P *trans* to anion	P *trans* to	
PEt_2Ph	111.1	84.0	31
$PPr^n{}_2Ph$	110.9	84.2	31
$PBu^n{}_2Ph$	111.0	84.1	31
$PMePh_2$	114.5	86.0	31
$PEtPh_2$	114.0	84.6	31
$RhCl_2(SFO_3)(PMe_2Ph)_3$	118	83	126
$[Rh_2Cl_4(PMe_2Ph)_6](BF_4)_2$	119	83	126
$[Rh(ttp)(CO)]PF_6$	115.2	112.3	124
RhCl(ttp)CO	115.5	114.7	124
$[Rh(NO)Cl(ttp)]PF_6$	138.9	105.8	124
$[Rh(CH_3)Cl(ttp)]FSO$	114.5	93.4	124
$[Rh(C_2H_5)Cl(ttp)]PF_6$	120.0	98.0	124
$[RhHCl(ttp)(MeCN)]BF_4$	116.2	89.0	124
$[RhHCl(ttp)(EtOH)]BF_4$	115.7	89.4	124
$[RhHCl(ttp)(CO)]PF_6$	89.7	81.3	124
$[Rh(CH_3)Cl(ttp)(MeCN)]FSO_3$	116.2	91.5	124
$[Rh(CH_3)Cl(ttp)(CO)]FSO_3$	108.1	83.8	124
$RhCl_3(CO)(PEtPh_2)_2$		74	117c
$RhCl_3(CO)(PClPh_2)_2$		86	117c
$RhHCl_2(CO)(PEtPh_2)_2$		81	117c
$RhH_2Cl(P(p\text{-}Tol)_3)_3$	114	90	116b
$RhH_2Cl(PPh_3)_3$	114	90	116b
$RhH_2Cl(PCy_3)_2$		113.7	127
$RhHCl_2(PCy_3)_2$		96.2	127
$RhH_2(BH_4)(PCy_3)_2$		108.3	127
$RhCl_2(COCH_2CH_2Ph)(PPh_3)_2$		108	122

Table 16. (continued)

Compound	$^{1}J(Rh,P)$, Hz		Ref.
	P *trans* to anion	P *trans* to	
$RhBr_2(COCH_2CH_2Ph)(PPh_3)_2$		106	122
$RhClBr(COCH_2ch_2Ph)(PPh_3)_2$		107	122
$RhCl_2(COMe)(PPh_3)_2$			123
"*cis*"	145		
"*trans*"		108	
$RhCl_2(COMe)(P(p\text{-}Tol)_3)_2$			123
"*cis*"	146		
"*trans*"		107	
$RhCl_2(COMe)(P(p\text{-}RPh)_3)_2$			123
"*cis*"	146		
"*trans*"		108	
$RhCl_2Me(CO)(PPh_3)_2$		90	123
$RhCl_2Me(CO)(P(p\text{-}Tol)_3)_2$		89	123
$RhCl_2Me(CO)(P(p\text{-}FPh)_3)_2$		91	123

Table 17. Two-Bond Couplings to Phosphorus as a Function of Complex Geometry

Complex	X	$^{2}J(P,X)$ *cis*,Hz	$^{2}J(P,X)$ *trans*,HZ
1. $[PtH(PEt_3)_3]ClO_4$ [a)]	^{1}H	- 17.5	+ 158.5
2. *cis*-$PtMe_2(PMe_2Ph)_2$ [b)]	^{13}C	9	104
3. *cis*-$Pt(^{15}NO_2)(PBu^n_3)_2$ [c)]	^{15}N	3	61
4. $[PtF(PEt_3)_3]BF_4$ [d)]	^{19}F	31.6	140
5. *cis*- and *trans*-$PdCl_2(PMe_3)_2$ [e)]	^{31}P	- 8.0	+ 610
6. *cis*-$PtCl(SnCl_3)(DIOP)$ [f)]	$^{117,119}Sn$	214	4,188
7. XXXXI, p. 59	^{199}Hg	397	3,879

Footnotes Table 17:

a) Ref. [*83*].
b) Cheney, A.J., Mann B.E. and Shaw, B.L.: Chem.Commun. (1971), *431*.
c) Kunz, R.W. and Pregosin, P.S. unpublished results.
d) Ref. [*179*].
e) Ref. [*148*].
f) Ref. [*210*].

Table 18. a) $^{2}J(PP')$ values in *cis*-disubstituted complexes

	$^{2}J(PP')$ (Hz)
$Cr(CO)_4(PH_3)_2$	- 26.2
$[P(SCH_2)_3CMe]_2$	~ 0
$(PMe_3)_2$	- 36 ± 1
$(PH_3)(PPh_3)$	33.0
$(PBu_3)(PH_3)$	38.5
$[P(OCH_2)_3CMe]_2$	70 ± 10
$[P(NMe_2)F_2]_2$	62 ± 2
$[P(CCl_2H)F_2]_2$	67 ± 0.5
$[P(CCl_3)F_2]_2$	67 ± 1
$(PF_3)_2$	78 ± 1,77.0
$[EtN(PF_2)_2]$	78.5 ± 0.5
$Mo(CO)_4(PH_3)_2$	18.3 - 18.9
$(PMe_3)_2$	- 29.7 ± 0.1
$(PBu_2Ph)(PPh_3)$	21
$(PPh_2H)_2$	~ 19
$(P(NMe_2)_3)_2$	12.4 ± 0.2
$[P(NMePh)_3]_2$	18.2 ± 0.2
$[P(SCH_2)_3CMe]_2$	30 ± 10
$[P(SMe)_3]_2$	35 ± 10
$[P(NMe_2)_3][P(OCH_2)_3CEt]$	39 ± 3

Table 18. a) (continued)

	$^2J(PP')$ (Hz)
$Mo(CO)_4[P(OMe)_3]_2$	- 40.5 ± 1
$[P(OMe)_3][P(OCH_2)_3CEt]$	48 ± 3
$[P(OCH_2)_3C\text{-}n\text{-}Pr]_2$	50 ± 10
$[P(NMe_2)F_2]_2$	38 ± 2
$[P(NEt_2)F_2]_2$	38 ± 2
$[P(CF_3)_2F]_2$	41.0 ± 1
$[P(CH_2Cl)F_2]_2$	42.5 ± 0.5
$[P(CCl_2H)F_2]_2$	47.5 ± 0.5
$[P(CCl_3)F_2]_2$	- 48.0 ± 0.5
$[P(CF_3)F_2]_2$	48.5 ± 1
$(PF_3)_2$	55.0 ± 0.2
$[EtN(PF_2)_2]$	+ 121.0
$W(CO)_4(PH_3)_2$	13.4
$[P(SCH_2)_3CMe]_2$	0
$[P(SMe)_3]_2$	0
$[PMe_3]_2$	- 25.0 ± 0.1
$[P(OCH_2)_3C\text{-}n\text{-}Pr]_2$	35 ± 5
$[P(NMe_2)F_2]_2$	21.1 ± 1.0
$[P(CH_2Cl)F_2]_2$	34.6 ± 0.8
$[P(CCl_2H)F_2]_2$	37.0 ± 0.5
$[P(CCl_3)F_2]_2$	37.6 ± 0.8
$(PF_3)_2$	38 ± 1,41
$[EtN(PF_2)_2]$	155.3 ± 0.6
$PdCl_2(PMe_3)_2$	8
$Br_2(PMe_3)_2$	1.0 ± 1
$Cl_2(PMe_2H)_2$	0 ± 1
$Br_2(PMe_2H)_2$	2
$Cl_2[P(OMe)_3]_2$	+ 79.9 ± 0.2

Table 18. a) (continued)

	$^2J(PP')$ (Hz)
$Cl_2[P(OCH_2)_3C$-n-Pent$]_2$	65 ± 10
$PtCl_2(PMe_2H)_2$	0 ± 1
$Br_2(PMe_2H)_2$	0
$I_2(PMe_2H)_2$	0
$Cl_2[P(NMe_2)_3]_2$	~ 0
$Cl_2(PMe_3)_2$	< 5, 18.9
$Br_2(PMe_3)_2$	16.2
$I_2(PMe_3)_2$	14.0
$Cl_2[P(OMe)_3]_2$	10 ± 2, 18.3 ± 0.1
$Br_2[P(OMe)_3]_2$	8.8 ± 0.1
$I_2[P(OMe)_3]_2$	~ 0,5.0 ± 0.1
$Cl_2[P(OCH_2)_3C$-n-Pent$]_2$	35 ± 10
$I_2[P(OCH_2)_3C$-n-Pent$]_2$	14 ± 5
$Cl_2[P(OPh)_3][PBu_3]$	20.0
$Cl_2[P(NR_2)_2F]_2$	25 - 37
$MeCl[P(OPh)_3][PBu_3]$	23.3

Table 18. b) $^2J(P,P)$ values for *trans*-disubstituted complexes (Data from Ref. [*92*])

	$^2J(P,P)$ (Hz)
$Cr(CO)_4[P(SCH_2)_3CMe]_2$	0
$[P(SMe)_3]_2$	0
$(PMe_3)_2$	- 28.5 ± 1
$(PBu_3)(PPh_3)$	25
$(PBu_3)[P(OPh)_3]$	30
$[P(NMe_2)_3]_2$	- 17 ± 5

Table 18. b) (continued)

	$^{2}J(P,P)$ (Hz)
$[P(OMe)_3]_2$	- 15.0 ± 0.1
$[P(OCH_2)_3CEt]_2$	9 ± 3
$(PF_3)_2$	34 ± 1,34.0 ± 0.5
$Mo(CO)_4(PBu_3)(PPh_3)$	50
$(PBu_2Ph)(PPh_3)$	49
$(PBu_3)[P(OPh)_3]$	112
$[P(NMe_2)_3]_2$	+ 101 ± 1
$[P(NMe_2)_3][P(OCH_2)_3CEt]$	141 ± 3
$[P(OMe)_3]_2$	+ 162 ± 5
$[P(OMe)_3][P(OCH_2)_3CEt]$	185 ± 3
$[P(OCH_2)_3C\text{-n-Pr}]_2$	210 ± 30
$(PF_3)_2$	312 ± 1
$W(CO)_4(PBu_3)(PPh_3)$	65
$(PBu_3)[P(OPh)_3]$	120
$(PBu_2Ph)[P(OPh)_3]$	112
$[P(NMe_2)_3]_2$	+ 81 ± 5
$[P(OCH_2)_3C\text{-n-Pr}]_2$	140 ± 25
$(PF_3)_2$	315 ± 1
$Fe(CO)_3[P(NMe_2)_3]_2$	+ 65 ± 10
$[P(NMe_2)_3][P(OCH_2)_3C\text{-n-Pr}]$	183 ± 1
$[P(OCH_2)_3CEt]_2$	300 ± 40
$PdCl_2(PMe_2H)_2$	515
$Br_2(PMe_2H)_2$	516
$PdI_2(PMe_3)_2$	+ 572 ± 5
$I_2(PMe_3)(PEt_3)$	565
$I_2(PBu_3)(PMe_2Ph)$	551
$I_2(PBu_3)[P(OPh)_3]$	758

Table 18. b) (continued)

	$^{2}J(P,P)$ (Hz)
$I_2(PMe_2Ph)[P(OPh)_3]$	829
$Cl_2[PPh_2(CF_3)]_2$	1100 ± 50
$PtCl_2(PMe_2H)_2$	515
$Br_2(PMe_2H)_2$	462
$I_2(PMe_2H)_2$	447
$Cl_2(PEt_3)_2$	90
$[P(OPh)_3][PBu_3]$	709
mer-$Mo(CO)_3(PEt_3)_3$	15
$[P(OMe)_3]_3$	47
$[P(OEt)_3]_3$	51
fac-$Mo(CO)_3(PPhF_2)_3$	- 43.0 ± 1.0
$P(OPh)_2F$	- 51.0 ± 1.6
$P(OPh)F_2$	- 55.2 ± 1.0
$(PF_3)_3$	- 56.0 ± 0.5
fac-$Mn(PhCO)(CO)_3[P(OCH_2)_3CEt]_2$	5 ± 2
$H(CO)_3(PF_3)_3$	125 ± 1
$Br(CO)_3[P(OMe)_3]_2$	140 ± 20
$Br(CO)_3[P(OCH_2)_3C\text{-n-Pr}]$	110 ± 20
sym-mer-$MnI(CO)_3[P(OMe)_3]_2$	170 ± 30
$H(CO)_3(PF_3)_2$	155 ± 1
$\pi\text{-}C_5H_5Mn(CO)[P(OCH_2)_3CMe]_2$	120 ± 20
$Fe(CO)_3(PF_3)_2$	98 ± 1
$RuCl_2(CO)_2(PMe_2H)_2$	308 ± 1
$RuCl_2(CO)_2(PMe_2H)_2$	229 ± 1
trans-$RhCl(CO)(PMe_2H)_2$	321 ± 1
$Br(CO)(PMe_2H)_2$	329 ± 1
$I(CO)(PMe_2H)_2$	337 ± 1

Table 18. b) (continued)

	$^{2}J(P,P)$ (Hz)
$RhCl_3(CO)(PMe_2H)_2$	495 $\pm$ 1
mer-$RhCl_3(PPr_3)_3$	22 $\pm$ 3
$(PBu_3)_3$	21 $\pm$ 3
$[PEt_2Ph]_3$	30 $\pm$ 8
sym-mer-$RhCl_3(PBu_3)_2[P(OPh)_3]$	30.3
trans-$IrCl_3(CO)(PMe_2H)_2$	386 $\pm$ 1
$Br_3(CO)(PMe_2H)_2$	390 $\pm$ 1
trans-$NiI_2(PMe_2H)_2$	355
$Ni(CO)_2[P(NMe_2)_3]_2$	~ 0
$Ni(CO)_2[P(OMe)_3]_2$	10 $\pm$ 3
$Ni(CO)_2[P(OCH_2)_3CMe]_2$	~ 0
$Ni(CO)_2[P(CF_3)_3]_2$	~ 0
$Ni(CO)_2[P(NMe_2)F_2]_2$	~ 0
$Ni(CO)_2[P(NEt_2)F_2]_2$	0
$Ni(CO)_2(PF_3)_2$	38 $\pm$ 1
$Ni[P(CF_3)_2F]_4$	5
$Ni(C_6H_4O_2PF)_4$	17.7 $\pm$ 0.2
$Ni[P(CH_2Cl)F_2]_4$	12 $\pm$ 5
$Ni[P(CCl_3)F_2]_4$	21 $\pm$ 1
$Ni[P(NMe_2)F_2]_4$	10
$Ni[P(CF_3)F_2]_4$	5
$Ni(PF_3)_4$	3 $\pm$ 1
$PdCl[OP(OPh)_2][HOP(OPh)_2]PR_3$	600-630
$PtCl[OP(OPh)_2][HOP(OPh)_2]PR_3$	539-547
	33.0-35.8
	23.4-23.9
$PtCl[P(OPh)_3[(PBu_3)_2]Cl$	24.0

Table 19. Values of $^{2}J(P,P)$ in some complexes

Complex	X: Cl	Br	I	CN
cis				
$[PdX_2(PMe_3)_2]$	- 8.0	ca. - 1		
$[PdX_2\{P(OMe)_3\}_2]$	+ 77.9	+ 84.8	+ 61.9	
$[PtX_2(PMe_3)_2]$	- 18.9	- 16.2	- 14.0	
$[PtX_2\{P(OMe)_3\}_2]$	- 17.8	- 8.4	ca. - 1.9	
trans				
$[PdX_2(PMe_3)_2]$	+ 610	+ 612	+ 604	+ 441
$[PdX_2(PEt_3)_2]$	+ 541	+ 537	+ 520	
$[PdX_2\{P(OMe)_3\}_2]$			+ 1,145	
$[PtX_2(PMe_3)_2]$	+ 510	+ 514	+ 498	+ 383
$[PtX_2(PEt_3)_2]$	+ 436	+ 449	+ 438	
$[PtX_2(PPr^n_3)_2]$		451		
$[PtX_4(PMe_3)_2]$	+ 660	+ 645		
$[PtX_4(PEt_3)_2]$	+ 587	+ 574		
$(Pr^n_4N)[IrX_4(PMe_3)_2]$	+ 539			
$[Hg(PMe_3)_2](NO_3)_2$			+ 250	
$[Au(PMe_3)_2][Au(CN)_2]$			+ 353	

tris, tetrakis

	Cl		Br	
	cis	*trans*	*cis*	*trans*
$[PdX(PMe_3)_3]NO_3$	- 25	+ 461	- 23	+ 469
$[PtX(PMe_3)_3]^+$	- 25	+ 414	- 23	+ 415
mer-$[PtX_3(PMe_3)_3]^+$	- 12	+ 556	- 13	+ 547
mer-$[RhX_3(PMe_3)_3]$	- 29	+ 567	28	+ 555
mer-$[IrX_3(PMe_3)_3]$	- 22	+ 452		
trans-$[RhX_2(PMe_3)_4]BF_4$	- 35	+ 469	- 35	+ 461
trans-$[IrX_2(PMe_3)_4]NO_3$	- 28	+ 374	- 27	+ 362

Table 19. (continued)

Complex	$^2J_{PP}$(*cis*)	$^2J_{PP}$(*trans*)
$[Pt(PMe_3)_4](NO_3)_2$	- 30	+ 335
$[Pt\{P(OMe)_3\}_4]I_2$	- 44	+ 689

Data from Ref. [*148*].

Table 20. ^{31}P data for some Rh(I) complexes

	δ P^a	δ P	$^2J(P,P)$	Ref.
$RhCl(PPh_3)_3$	48.0	31.5	- 38	116a
$RhBr(PPh_3)_3$	46.8	29.8	- 37	116a
$RhI(PPh_3)_3$	43.2	27.1	- 36	116a
trans-$RhCl(CO)(PMe_2Ph)_2$	- 2.0		+ 376	113
trans-$RhBr(CO)(PMe_2Ph)_2$	- 4.1		+ 368	113
trans-$RhI(CO)(PMe_2Ph)_2$	- 7.2		+ 361	113
trans-$RhCl(CO)(PRBu^t_2)_2$ [b]				
<u>R</u>				
Me	{46.6, 30.7}		317.0	118b
Et	{58.2, 47.2}		306.0	118b
Pr	{55.9, 44.5}		307.6	118b
$RhCl(CO)(PMePh_2)(PPh_3)$			361.8	117a
$RhCl(CO)(PMePh_2)(PEt_2Ph_2)$			352.4	117a
$RhCl(CO)(PMePh_2)(PEt_2Ph)$			348.3	117a
$RhCl(CO)(PMePh_2)(PEt_3)$			344.3	117a
$RhCl(CO)(PEt_2Ph)(P(OPh)_3)$	119.0, 26.3		537.8	117a

Table 20. (continued)

	δ P[a]	δ P	$^{2}J(P,P)$	Ref.
RhCl(CO)(dppe)	63.4[c]	42.2[c]	34	117c
RhBr(CO)(dppe)	61.6[c]	40.3[c]	33	117c
$RhCl(CO)(PEtPh_2)_2$	85.9[c]			117c

a) *trans* to halogen.

b) Give ABX ^{31}P spectra due to different rotamers in solution.

c) ppm upfield from P_4O_6.

Table 21. ^{31}P NMR data for some iridium(III) complexes

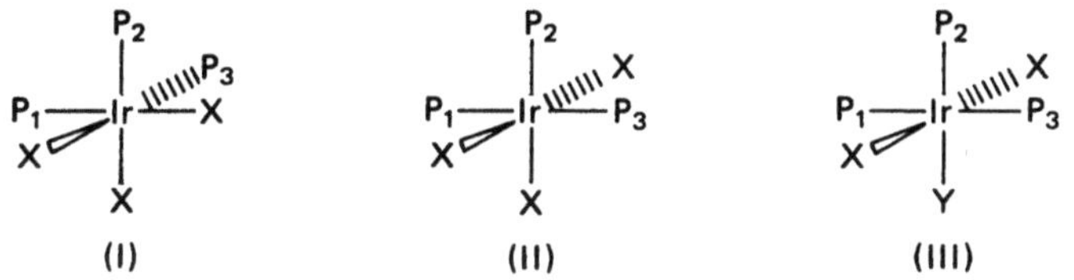

L	X	Configuration	δP_2	δP_3	$^{2}J(P_1P_2)$	Ref.
PMe_3	Cl	(II)	- 47.0	- 42.4	19.5	31
PMe_2Ph	Cl	(II)	- 50.1	- 40.1	17.1	31
PMe_2Ph	Br	(II)	- 54.6	- 52.5	16.5	31
PMe_2Ph	I	(II)	- 64.7	- 72.4	16.3	31
PMe_2Ph	Br	(III)	- 49.6	- 45.1	17.0	31
PMe_2Ph	I	(III)	- 52.6	- 51.7	17.2	31
PMe_2Ph	Me	(III)	- 43.8	- 54.2	17.1	31
PBu^n_3	Cl	(II)	- 40.3	- 37.7	16.5	31
PBu^n_3	Br	(II)	- 42.6	- 47.8	15.7	31
PBu^n_3	I	(II)	- 49.1	- 63.3	15.0	31
PBu^n_3	Br	(III)	- 39.6	- 43.1	15.5	31

Table 21. (continued)

L	X	Configuration	δP_2	δP_3	$^2J(P_1P_2)$	Ref.
	Cl	III	- 39.6	5.2	450	182
	Br	III	- 43.3	4.4	447	182
	I	III	- 50.0	1.3	438	182
	NO_2	III	- 36.3	12.5	408	182
	py	III	- 34.6	10.7	398	182
	CO	III	- 41.0	19.2	320	182
PEt_3	Cl	I	- 26.2			30
PPr_3	Cl	I	- 31.9			30
PEt_2Ph	Cl	I	- 34.8			30
PPr_2Ph	Cl	I	- 39.3			30
PBu_2Ph	Cl	I	- 39.4			30
$PEtPh_2$	Cl	I	- 37.3			30
$PPrPh_2$	Cl	I	- 38.6			30

Table 21. (continued)

L	X	Configuration	δP_2	δP_3	$^2J(P_1P_3)$	Ref.
$PBu^n{}_3$	I	(III)	- 42.5	- 47.4	15.4	31
$PMePh_2$	Cl	(II)	- 55.2	- 47.1	15.6	31
$PMePh_2$	Br	(III)	- 56.9	- 47.3	15.4	31
$PMePh_2$	I	(III)	- 63.6	- 56.1	14.7	31
$PPr^n{}_2Ph$	Cl	(II)	- 41.6	- 38.3	15.9	31
PEt_3	Cl	(II)	- 36.1	- 33.3	16.5	31
PEt_3	Br	(II)	- 38.3	- 42.8	15.8	31
PEt_3	I	(II)	- 43.5	- 56.9	15.1	31
PEt_3	H	(II)	- 8.9	- 0.1	19	31
PEt_3	Br	(III)	- 35.2	- 37.3	15.5	31
PEt_3	I	(III)	- 38.7	- 43.1	15.5	31
PEt_2Ph	Cl	(II)	- 37.5	- 33.9	15.8	31
PEt_2Ph	Br	(II)	- 41.5	- 44.6	15.6	31
PEt_2Ph	I	(II)	- 50.5	- 60.3	15.3	31
PEt_2Ph	Br	(III)	- 37.4	- 37.4		31
PEt_2Ph	I	(III)	- 41.2	- 42.6	15.1	31
PEt_2Ph	NO_3	(III)	- 38.9	- 23.4	15.7	31

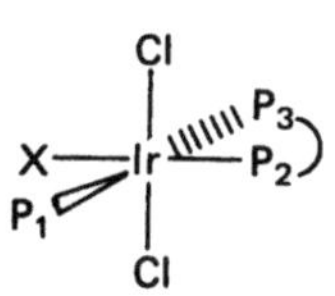

			δP_2	δP_3	$^2J(P_1P_3)$	Ref.
$(PMe_2Ph)(diphos)$						
	NO_3	III	- 29.9	17.3	432	182
	N_3	III	- 34.5	8.3	445	182
	-NCO	III	- 36.5	10.9	434	182
	SCN-	III	- 42.1	4.2	441	182

Table 22. ^{31}P chemical shifts[a] for some Cr(O), Mo(O) and W(O) complexes

$Cr(CO)_5P$		$Mo(CO)_5P$		$W(CO)_5P$	
P	$\delta^{31}P$	P	$\delta^{31}P$	P	$\delta^{31}P(^1J(W,P))$
PBu_3	30.2	PBu_3	12.2	PBu_3	- 6.4 (200)
PPh_3	55.3	PPh_3	37.5	PPh_3	20.6 (280)
$PMePh_2$	35.0	$PMePh_2$	15.0	$PMePh_2$	- 3.8 (245)
$PEtPh_2$	48.3	$PEtPh_2$	30.1	$PEtPh_2$	12.1 (240)
$P\textit{i}PrPh_2$	59.5	$P\textit{i}PrPh_2$	43.2	$P\textit{i}PrPh_2$	26.3 (240)
$PBuPh_2$	45.1	$PBuPh_2$	26.9	$PBuPh_2$	7.9 (250)
$P\textit{t}\text{-}BuPh_2$	72.7	$P\textit{t}\text{-}BuPh_2$	57.0	$P\textit{t}\text{-}BuPh_2$	41.7 (240)
PBu_2Ph	35.3	PBu_2Ph	17.6	PBu_2Ph	- 0.8 (235)
		$P\textit{n}\text{-}PrPh_2$	26.5		

a) Data from Ref. [*40*]

Table 23. ^{31}P NMR data for $L_2M(CO)_4$

Compound	Chem. shift ppm	Ref.
trans-$((C_4H_9)_3P)_2Cr(CO)_4$	+ 44.2	106
cis-$(C_6H_5(C_4H_9)_2P)_2Cr(CO)_4$	+ 32.4	106
trans-$(C_6H_5(C_4H_9)_2P)_2Cr(CO)_4$	+ 50.4	106
trans-$((C_6H_5)_2C_4H_9P)_2Cr(CO)_4$	+ 61.5	106
cis-$((C_4H_9)_3P)_2Mo(CO)_4$	+ 8.9	106
trans-$((C_4H_9)_3P)_2Mo(CO)_4$	+ 22.4	106
cis-$(C_6H_5(C_4H_9)_2P)_2Mo(CO)_4$	+ 14.5	106
trans-$(C_6H_5(C_4H_9)_2P)_2Mo(CO)_4$	+ 28.8	106
cis-$((C_6H_5)_2C_4H_9P)_2Mo(CO)_4$	+ 25.8	106
trans-$((C_6H_5)_2C_4H_9P)_2Mo(CO)_4$	+ 38.9	106
cis-$((C_4H_9)_3P)_2W(CO)_4$	- 10.0	106
trans-$((C_4H_9)_3P)_2W(CO)_4$	- 2.5	106
cis-$(C_6H_5(C_4H_9)_2P)_2W(CO)_4$	- 4.0	106
trans-$(C_6H_5(C_4H_9)_2P)_2W(CO)_4$	+ 4.0	106
cis-$((C_6H_5)_2C_4H_9P)_2W(CO)_4$	+ 8.2	106
trans-$((C_6H_5)_2C_4H_9P)_2W(CO)_4$	+ 14.4	106

Table 24. ^{31}P values for some complexes of palladium

trans-$PdCl_2P_2$		
P	$\delta^{31}P$	Ref.
PEt_3	17.8	194
PPr_3	9.2	194
$PBu^n{}_3$	10.0	194
PMe_2Ph	- 5.0, - 5.2	194, 200
PEt_2Ph	16.1, 15.7	194, 200
PPr_2Ph	10.2	194

Table 24. (continued)

trans-$PdCl_2P_2$ P	$\delta^{31}P$	Ref.
$PBu^n{}_2Ph$	10.9	194
$PMePh_2$	7.8	194
$PEtPh_2$	19.3, 19.1	194, 200
$PPrPh_2$	16.3	194
PBu^nPh_2	16.3	194
$P(p\text{-}Tol)_3$	21.4	194
PMe_2Bu^t	10.7	194
PEt_2Bu^t	32.1	194
PPr_2Bu^t	26.8	194
$PBu^n{}_2Bu^t$	27.4	194
$P(p\text{-}Tol)_2Bu^t$	38.0	194
PPh_2Bu^t	39.2	194
$PPrBu^t{}_2$	40.7	194
$P(p\text{-}Tol)Bu^t_2$	52.3	194
$PPhBu^t{}_2$	52.9	194
$PMe_2(p\text{-}OCH_3C_6H_4)$	- 6.2	198
$PMe_2(p\text{-}ClC_6H_4)$	- 4.9	198
$PMe_2(p\text{-}CH_3C_6H_4)$	- 5.7	198
$PMe(p\text{-}ClC_6H_4)_2$	7.4	198
$PMe(p\text{-}CH_3C_6H_4)_2$	6.3	198
PBz_3	13.3	198
PBz_2Ph	14.2	198
$PBzPh_2$	20.1	198

Table 25. Phosphorus chemical shifts in the compounds *trans*-[$PdCl_2L$(phosphine)]

		$\delta^{31}P$	Ref.
PBu_3^n	L		
	PBu_3^n	10.0	196
	$AsBu_3^n$	13.2	196
	$PTol_3$	16.2	196
	$AsTol_3$	20.4	196
	$NHEt_2$	24.3	196
	pyridine	27.3	196
$PTol_3$	L		
	PBu_3^n	17.9	196
	$AsBu_3^n$	18.6	196
	$PTol_3$	21.4	196
	$AsTol_3$	23.0	196
	$NHEt_2$	25.2	196
	pyridine	27.3	196
sym.-trans	[$Pd_2Cl_4(PBu_3^n)_2$]	39.7	196
sym.-trans	[$Pd_2Cl_4(PTol_3)_2$]	31.5	196

Table 26. ^{31}P NMR chemical shifts[a] (ppm) for some silver(I) and mercury(II) complexes

		$\delta^{31}P$	Ref.
1	$Ag(p\text{-}Tol_3P)_2PF_6$[b]	33.4[c]	33
2	$Ag(p\text{-}Tol_3P)_2NO_3$[b]	29.8[c]	33
3	$Ag(p\text{-}Tol_3P)_2Cl$[b]	21.4[c]	33
4	$Ag(p\text{-}Tol_3P)_3PF_6$[b]	30.6[c]	33
5	$Ag(p\text{-}Tol_3P)_3NO_3$[b]	27.3[c]	33

Table 26. (continued)

		$\delta^{31}P$	Ref.
6	$Ag(p\text{-}Tol_3P)_3Cl$ [b]	22.7 [c]	33
7	$Ag(p\text{-}Tol_3P)_3I$ [b]	18.8 [c]	33
8	$Ag(p\text{-}Tol_3P)_4PF_6$ [b]	25.6 [c]	33
9	$Ag(n\text{-}Bu_3P)_2BF_4$ [d]	32.9 [e]	95
10	$Ag(n\text{-}Bu_3P)_3BF_4$ [d]	26.6 [e]	95
11	$Ag(n\text{-}Bu_3P)_4BF_4$ [d]	15.5 [e]	95
12	$Ag(\underline{\underline{1}})BF_4$ [f]	12.1	97
13	$Ag(\underline{\underline{1}})NO_3$ [f]	8.0	97
14	$Ag(\underline{\underline{1}})Cl$ [f]	6.3	97
15	$Ag(\underline{\underline{1}})I$ [f]	2.6	97
16	$Ag(\underline{\underline{2}})BF_4$ [f]	28.5	205
17	$Ag(\underline{\underline{2}})NO_3$ [f]	24.0	205
18	$Ag(\underline{\underline{2}})Cl$ [f]	20.9	205
19	$Ag(\underline{\underline{2}})I$ [f]	20.9	205
20	$Ag(\underline{\underline{3}})BF_4$ [f]	59.1	205
21	$Ag(\underline{\underline{3}})NO_3$ [f]	53.2	205
22	$Ag(\underline{\underline{3}})Cl$ [f]	53.5	205
23	$Ag(\underline{\underline{3}})I$ [f]	50.1	205
24	$Ag(\underline{\underline{1}})\{P(OCH_2)_3CEt)\}BF_4$ [f]	9.1	205
25	$Ag(\underline{\underline{1}})\{P(OCH_2)_3CEt\}_2BF_4$ [f]	3.1	205
26	$Hg(Ph_2PCH_2Ph)_2(NO_3)_2$	49.7	82
27	$Hg(Ph_2PCH_2Ph)_2Cl_2$	32.0	82
28	$Hg(Ph_2PCH_2Ph)_2Br_2$	25.9	82
29	$Hg(Ph_2PCH_2Ph)_2I_2$	11.9	82
30	$Hg(PSP)Br_2$ [g]	12.2	82
31	$Hg(PSP)I_2$ [g]	- 4.5	82
32	$Hg(Cy_3P)_2(NO_3)_2$	66.7	82
33	$Hg(Cy_3P)_2Cl_2$	52.5	82

Table 26. (continued)

		$\delta^{31}P$	Ref.
34	$Hg(Cy_3P)_2Br_2$	46.1	82
35	$Hg(Cy_3P)_2I_2$	35.6	82
36	$Hg(n\text{-}Oc_3P)_2Cl_2$	29.0	49
37	$Hg(n\text{-}Oc_3P)_2Br_2$	22.5	49
38	$Hg(n\text{-}Oc_3P)_2I_2$	10.0	49
39	$Hg(n\text{-}Bu_2PPh)_2Cl_2$	28.6	50
40	$Hg(n\text{-}Bu_2PPh)Br_2$	22.1	50
41	$Hg(n\text{-}Bu_2PPh)I_2$	6.1	50

a) Relative to H_3PO_4 unless otherwise stated. A positive sign indicates a shift to lower field.

b) Data obtained at - 80° in CH_2Cl_2 solution.

c) Values reported with reference to an internal capillary of PEt_3.

d) At - 80°.

e) Relative to external PBu_3^n.

f) (1) is 2,11 bis(diphenylphosphinomethyl)benzo[c]phenenthrene..
(2) is 2,11 bis(dicyclohexylphosphinomethyl)benzo[c]phenanthrene.
(3) is 2,11 bis(dit-butylphosphinomethyl)benzo[c]phenanthrene.

g) PSP = $(Ph_2PCH_2CH_2)_2S$.

Table 27. ^{31}P Chemical shifts (p.p.m.) of the complexes PdL_n (L = tertiary phosphine) measured in toluene containing 10% $CD_3C_6D_5$ at - 60° C except where otherwise stated. Chemical shifts are quoted with respect to 85% H_3PO_4 with the sign convention that increasing frequency is positive

L	PdL_2	PdL_3	PdL_4	Ref.
PMe_3			- 34.8	199
PMe_2Ph			- 20.8	199
$PMePh_2$			- 4.2	199
PPh_3		+ 22.6[a)]	+ 18.4[a)]	199
PEt_3		+ 9.6	- 1.5	199

Table 27. (continued)

L	PdL_2	PdL_3	PdL_4	Ref.
PBu^n_3		$- 1.4^{b)}$	$- 7.9^{b)}$	199
$P(CH_2Ph)_3$		+ 2.7		199
PPr^i_3	$+ 49.3^{c)}$	$+ 39.0^{c)}$		199
$P(C_6H_{11})_3$	+ 38.7	+ 25.9		199
PBu^t_2Ph	+ 67.0			199

a-c) Measured at - 70, - 100, and - 80°C.

Table 28. ^{31}P data for transition metal-phosphorus chelate complexes

		δ_f a)	δ_c a)	Δ b)	Δ_R c)	chelate ring size	$^1J(M,P)$ Hz
Cr(O)	Diphosphines						
	$Cr(CO)_4(Ph_2PCH_2PPh_2)$	+ 23.6	- 25.4	- 49.0	+ 12	4	
	$Cr(CO)_4(Ph_2PCH_2CH_2PPh_2)$	+ 12.5	- 79.4	- 91.9	- 31	5	
	$Cr(CO)_4(Ph_2PCH_2CH_2CH_2PPh_2)$	+ 17.3	- 41.4	- 58.7	+ 2	6	
	cis-$Cr(CO)_4(Ph_2P\text{-n-Bu})_2$	+ 17.1	ca. - 44	ca. - 61			
Mo(O)	Diphosphines						
	$Mo(CO)_4(Ph_2PCH_2PPh_2)$	+ 23.6	0.0	- 23.6	+ 19.3	4	
	$Mo(CO)_4(Ph_2PCH_2CH_2PPh_2)$	+ 12.5	- 54.7	- 67.2	- 24.3	5	
	$Mo(CO)_4(Ph_2PCH_2CH_2CH_2PPh_2)$	+ 17.3	- 21.1	- 38.3	+ 4.6	6	
	cis-$Mo(CO)_4(PPh_2\text{-n-Bu})_2$	+ 17.1	- 25.8	- 42.9			
	Alkenylphosphines						
	$Mo(CO)_4(PPh(CH_2CH_2CH{=}CH_2)_2)$	+ 26.2	- 42.5,-44.3 e)	- 67.9,-69.0	- 28.6	5.5	
	cis-$Mo(CO)_4(PPh(n\text{-}Bu)_2)_2$	+ 26.2	- 14.5	- 40.7			
W(O)	Diphosphines						
	$W(CO)_4(Ph_2PCH_2PPh_2)$	+ 23.6	+ 23.6	0	+ 25.3	4	202
	$W(CO)_4(Ph_2PCH_2CH_2PPh_2)$	+ 12.5	- 40.1	- 52.6	- 27.3	5	231

Table 28. (continued)

		δ_f a)	δ_c a)	Δ b)	Δ_R c)	chelate ring size	$^1J(M,P)$ Hz
	$W(CO)_4(Ph_2PCH_2CH_2CH_2PPh_2)$	+ 17.3	0.0	- 17.3	+ 8.0	6	222
	cis-$W(CO)_4(PPh_2\text{-n-Bu})_2$	+ 17.1	- 8.2	- 25.3			230
Ru(II)	Alkenylphosphines						
	$RuCl_2[PPh_2(o\text{-}CH_2{=}CHC_6H_4]_2$ f)	+ 6.8	- 47.2	- 5	- 25.6	5.5	
	$RuCl_2(CO)_2(PPh_2Et)_2$ g)	+ 12.5	- 15.9	- 27.9			
Rh(I)	Alkenylphosphines						
	$[RhCl(PPh_2(CH_2CH_2CH{=}CH_2))]_2$ h)	+ 17.1	- 69.6	- 86.7	- 30.4	5.5	188
	$[RhCl(P(p\text{-}CH_3C_6H_4)_3)]_2$ i)	+ 6.8	- 49.5	- 56.3			196
Ir(III)	Diphosphines						
	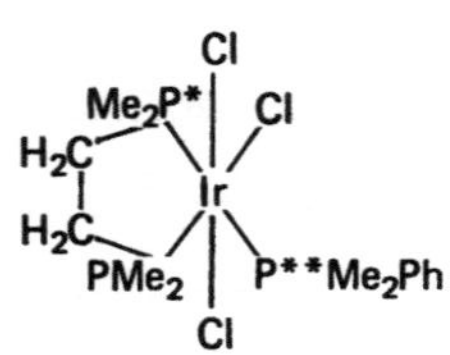	+ 47.6	- 4.2 (*)	- 51.8	- 46.3 j)	5	
			+ 42.1 (**)	- 5.5			
	Internally metallated phosphines						

	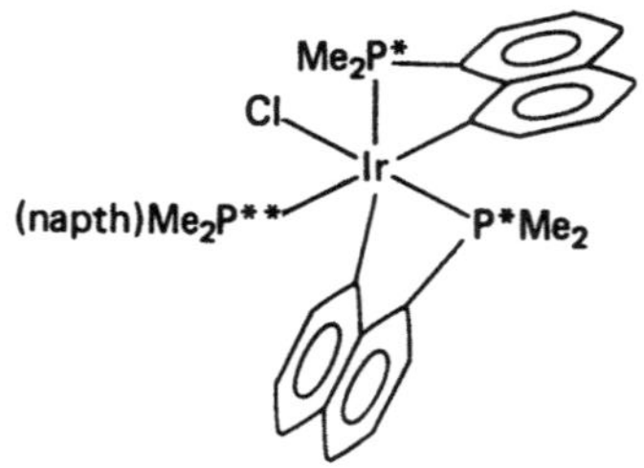	+ 47.6	+ 15.1 (*)	- 32.5	- 30.8	5
			+ 14.5 (*)	- 33.1	- 30.2	
			+ 45.6 (**)	- 1.95		
Ni(O)	Diphosphines					
	$Ni[Me_2PCH_2CH_2PMe_2]_2$	+ 49.4	- 18.1	- 67.5	- 26.4	5
	$Ni[PMe_3]_4$	+ 63.3	+ 22.2	- 41.1		
Pd(II)	Diphosphines					
	$PdCl_2(Ph_2P^*CH_2OPPh_2)$	+ 14.1	- 79.8	- 93.9	- 52.1	5
	cis-$PdCl_2(Ph_2P^*CH_2OPPh_2)_2$	+ 14.1	- 27.0	- 41.8		
	$PdCl_2(Ph_2P^*CH_2CH_2OPPh_2)$	+ 22.8	- 11.7	- 34.5	+ 12.3	6
	$PdCl_2(Ph_2P^*CH_2CH_2OPPh_2)_2$	+ 22.8	- 24.0	- 46.8		
	Alkenylphosphines					
	$PdCl_2(Ph_2PCH_2CH_2CH{=}CH_2)$	+ 17.1	- 57.1	- 74.2	- 30.5	5.5
	cis-$PdCl_2(Ph_2PCH_2CH_2CH{=}CH_2)_2$	+ 17.1	- 26.6	- 43.7		
	Internally metallated phosphines					
	$PdCl(PPh_3)$ $[(o\text{-}CH_2C_6H_4P\text{-}t\text{-}Bu(o\text{-}CH_3C_6H_4)]$ k)	- 15.9	- 58.8	- 42.9	- 20.8	5

Table 28. (continued)

	δ_f [a)]	δ_c [a)]	Δ [b)]	Δ_R [c)]	chelate ring size	$^1J(M,P)$ Hz
trans-$PdCl_2[P\text{-}t\text{-}Bu(p\text{-}CH_3C_6H_4)_2]_2$	- 15.9	- 38.0	- 22.1			
Pt(II) Alkenylphosphines						
$PtI_2(Ph_2PCH_2CH_2CH{=}CH_2)$	+ 17.1	- 34.1	- 51.2	- 27.1	5.5	3071
cis-$PtCl_2(Ph_2P\text{-}n\text{-}Bu)_2$	+ 17.1	- 7.0	- 24.1			3641
Internally metallated phosphines						
$PtCl[P^{**}(t\text{-}Bu)_2Ph]$ $[o\text{-}C_6H_4P^{*}(t\text{-}Bu)_2]$ [l)]	+ 37.99	+ 31.7 (*)	+ 69.7	+ 80.3 [m)]	4	2222
		- 48.6 (**)	- 10.61			2912
$PtCl[P^{**}\text{-}t\text{-}Bu(o\text{-}CH_3C_6H_4)_2]$	+ 37.99	- 52.0 (*)		- 20.9	5	3262
$[o\text{-}CH_2C_6H_4P\ \text{-}t\text{-}Bu(o\text{-}CH_3C_6H_4)]$		- 31.1 (**)				2755

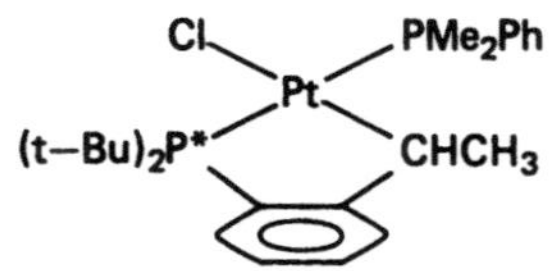	+ 37.99[n)]	- 73.0 (*)	- 35.0	- 31.8	5	3279
trans-$PtCl_2[P(t-Bu)_2Ph]_2$	+ 37.99	- 41.2	- 3.2			
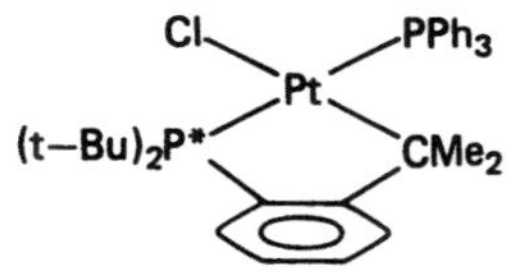	+ 37.99	- 71.3	- 33.3	- 30.1	5	
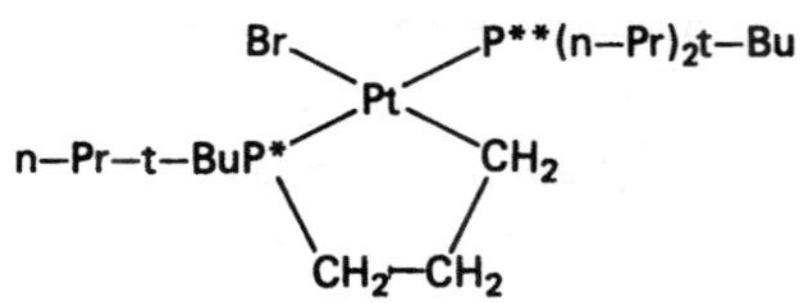	+ 8.74	- 67.4 (*)	- 76.1	- 41.5	5	3086
	+ 8.74	- 25.9 (**)	- 34.6			2984
trans-$PtCl_2[P(n-Pr)_2-t-Bu]_2$	+ 8.74	- 34.3	- 43.1			
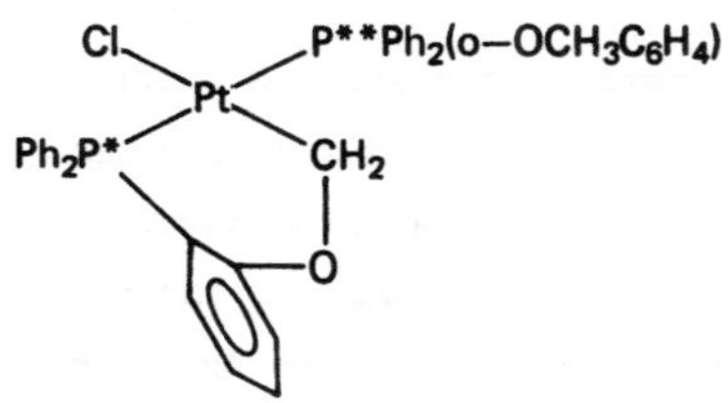	+ 6.8	- 8.3[p)]	- 15.1	+ 16.8	6	3178
	+ 6.8	- 25.1[p)]	- 31.9			3177

Footnotes Table 28.

a) In ppm downfield from 85% H_3PO_4, a negative sign indicates a downfield shift
b) $\Delta = \delta_1 - \delta_c$; δ_f = chemical shift of free ligand and δ_c = chemical shift of coordinated ligand.
c) $\Delta_R = \Delta(\text{obsd}) - \Delta(\text{coordinated for nonchelate})$.
d) Estimated by comparison to analogous molybdenum and tungsten complexes (see discussion).
e) Two isomers observed (see Ref. [*10*]).
f) Cl and olefin *cis*; P *trans*; δ_f given for $P(p\text{-}CH_3C_6H_4)_3$
g) Cl and CO *cis*; P *trans*
h) Cl bridged; P *trans* across dimer.
i) Cl bridged.
j) Data obtained by $^1H\{^{31}P\}$INDOR methods.
k) *Trans* P; δ_f given for $P(p\text{-}CH_3C_6H_4)_2\text{-t-Bu}$.
l) *Trans* P.
m) For $PtCl_2[PPh(t\text{-}Bu)_2]_2$ $_R$(calcd) = - 79.2 ppm.
n) δ_f given for $PPh(t\text{-}Bu)_2$.
o) New corrected assignment for isomer a.
p) δ_f given for $P(p\text{-}CH_3C_6H_4)_3$.

Table 29. ^{31}P data for some ruthenium(II) complexes

Compound	$\delta^{31}P$	$^{2}J(P,P)$	Ref.
trans-$RuCl_2(CO)_2L_2$			
<u>L</u>			
PMe_2Ph	- 2.90		206
$PBu^n{}_2Ph$	+ 10.62		206
PEt_3	+ 16.38		206
PEt_2Ph	+ 15.92		206
$PBu^tPr^n{}_2$	+ 27.40		206
$PBu^tBu^n{}_2$	+ 28.07		206
PBu^tEt_2	+ 32.41		206
cis-$RuCl_2(CO)_2L_2$			
<u>L</u>			
PMe_2Bu^t	+ 28.3		206
$PMePh_2$	+ 11.5		206
$PPr^n{}_2Ph$	+ 7.7		206
$PBu^n{}_2Ph$	+ 8.2		206
PEt_3	+ 15.7		206
PEt_2Ph	+ 13.0		206
$PEtPh_2$	+ 19.7		206
$PPr^n{}_2Bu^t$	+ 32.3		206
PPh_3	+ 17.4		206
$PBu^n{}_2Bu^t$	+ 32.7		206
PEt_2Bu^t	+ 36.1		206
$PBu^t(p\text{-}Tol)_2$	+ 28.5		206
$PBu^t{}_2Pr^n$	+ 30.2		206
$PBu^t{}_2Pr^n$	+ 45.6		206
$PHBu^t{}_2$	+ 53.2		206

Table 29. (continued)

Compound	$\delta^{31}P$	$^2J(P,P)$	Ref.
all *trans*-$RuCl_2(CO)_2(PHBu^t_2)_2$		229.0	206
cis-$RuCl_2(CO)_2(PHBu^t_2)_2$		308.0	206
$RuCl_2(PPh_3)_3$	75.7, 24.1	30.5	
$RuBr_2(PPh_3)_3$	80.0, 27.4	30.5	201
$RuHCl(PPh_3)_3$	94.0, 38.4	29.0	201
$RuHBr(PPh_3)_3$	93.8, 38.6	29.3	201
$[RuCl_2(PPh_3)_2]_2$	58.8, 53.0	41.5	201
$[RuBr_2(PPh_3)_2]_2$	61.6, 55.0	41.5	

Table 30. ^{31}P chemical shifts for some nickel(O) compounds

Compound	δ	Ref.	Compound	δ	Ref.
$Ni(CO)_3P(CH_3)_3$	- 19.9	207	$Ni(CO)_3(P(OCH_3)_3)$	161.4	207
$Ni(CO)_2(P(CH_3)_3)_2$	- 182.	207	$Ni(CO)_2(P(OCH_3)_3)_2$	165.1	207
$Ni(CO)(P(CH_3)_3)_3$	- 18.4	207	$Ni(CO)(P(OCH_3)_3)_3$	165.8	207
			$Ni(P(OCH_3)_3)_4$	162.3	207
$Ni(CO)_3(PCl_3)$	182.5	207	$Ni(CO)_3(PF_3)$	136.5	207
$Ni(CO)_2(PCl_3)_2$	179.4	207	$Ni(CO)_2(PF_3)_2$	136.8	207
$Ni(CO)(PCl_3)_3$	175.7	207	$Ni(CO)(PF_3)_3$	137.2	207
$Ni(PCl_3)_4$	172.7	207	$Ni(PF_3)_4$	137.7	207
$Ni(PMe_3)_4$	22.2	208	$Ni(P(O\text{-}pC_6H_4OCH_3)_3)_4$	131.6	208
$Ni(P(OEt)_2Ph)_4$	162.1	208	$Ni(P(O\text{-}pC_6H_4CH_3)_3)_4$	130.4	208
$Ni(P(OMe)_3)_4$	163.3	208	$Ni(P(O\text{-}o\text{-}C_6H_4CH_3)_3)_4$	129.0	208
$Ni(P(OEt)_3)_4$	58.4	208	$Ni(P(OPh)_3)_4$	128.8	208
$Ni(P(OCH_2CH_2Cl)_3)_4$	156.7	208			

Table 31. ^{13}C data[a] for olefins and (olefin)Ni[P(O-o-Tol)$_3$]$_2$ complexes

Olefin	$\delta(C_{free})$	$\delta(C_{coord})$	$\Delta\delta(C)$[b]	$^1J(C-H)_{free}$	$^1J(C-H)_{coord}$
Ethylene					
$H_2C=$	122.8	47.4	75.4	157	160 ± 5
Acrylonitrile[c]					
$H_2C=$	137.8	42.7	95.1	166	165 ± 5
=CH-	108.2	25.9	82.3	182	175 ± 5
-CN	117.5	e)			
Fumaronitrile[c]					
=CH-	117.8	24.8	93.0	184	172 ± 5
-CN	114.1	e)			
Methyl acrylate					
$CH_2=$	129.9	44.7	85.2	162	156[f]
=CH-	128.5	52.4	76.1	170	160[g]
$-CO_2-$	166.0	172.5	- 6.5		
$-CH_3$	50.8	50.2	0.6	146	146
Dimethyl maleate					
=CH-	130.0	51.1	78.9	168	160[h]
$-CO_2-$	165.6	170.7	- 5.1		

Table 31. (continued)

Olefin	$\delta(C_{free})$	$\delta(C_{coord})$	$\Delta\delta(C)$[b)]	$^1J(C\text{-}H)_{free}$	$^1J(C\text{-}H)_{coord}$
$-CH_3$	51.6	49.6	2.0	148	146
Maleic anhydride[d)]					
$=CH-$	137.1	47.8	89.3	188	160[i)]
$-CO_2-$	165.0	e)			

a) Chemical shifts of carbons in free and coordinated olefins in ppm (± 0) downfield of TMS. The uncertainty in $^1J(C\text{-}H)$ is $\pm$ 2 Hz except where noted otherwise.

b) Positive values of $\Delta\delta(C) = \delta(C_{free}) - \delta(C_{coord})$ indicate an upfield shift.

c) $^2J(C\text{-}H)$ values in the free olefin are 7 $\pm$ 1 Hz.

d) $^2J(C\text{-}H)$ = 4 Hz.

e) not observed.

f) $|J_{CP}+J_{CP'}|$ = 23 Hz.

g) $|J_{CP}+J_{CP'}|$ = 17 Hz.

h) $|J_{CP}+J_{CP'}|$ = 23 Hz.

i) $|J_{CP}=J_{CP'}|$ = 24 Hz.

Table 32. ^{13}C NMR chemical shifts for $L_nNi(CO)_{4-n}$ derivatives[a,b]

Ligand	$LNi(CO)_3$	$L_2Ni(CO)_2$	$L_3Ni(CO)$
PCl_3	190.2_3		
CO	191.6_4	191.6_4	191.6_4
$PPhCl_2$	192.3_3	192.9_4	
$P(OPh)_3$	193.3_3	194.7_9	
PPh_2Cl	194.1_3	196.2_8	
$PPh(OPh)_2$	194.1_8		
$P(OMe)_3$	194.8_2	197.9_9	200.7_6
$PPh_2(OPh)$	194.9_6		
$PPh(OMe)_2$	195.1_2		
$P(OEt)_3$	195.2_5	198.3_7	201.2_2
$P(O\text{-}i\text{-}Pr)_3$	195.5_4	198.9_4	
$PPh_2(OMe)$	195.6_0		
$AsPh_3$	195.8_0		
PPh_3	195.9_4	199.4	
PPh_2Me	196.1_7		
$P(p\text{-}Tol)_3$	196.2_1		
$PPhMe_2$	196.4_0		
$SbPh_3$	196.5_0		
PMe_3	196.6_9		
PEt_3	197.1_8	202.1_1	
PBu_3	197.3_3	202.0_4	
$(Ph_2PCH_2)_2$		200.7	

a) Chemical shifts in ppm downfield from TMS in $CDCl_3$ solution.
b) Data from Ref. [*237*].

Table 33. Some ^{13}C shieldings and coupling constants for the complexes *trans*-$[(CH_3)Pt(P(CH_3)_2(C_6H_5))_2L]^+PF_6^-$ a)

L	Platinum methyl		
	δC	$\lvert ^1J_{PtC} \rvert$	$\lvert ^2JPPtC \rvert$
$(CH_3)_2NCHO$	- 27.8	698 (±2)	5 (±2)
$p\text{-}CH_3OC_6H_4CN$	- 19.1	652 (±2)	6 (±2)
$CH_3C{\equiv}CCH_3$	- 4.2	632 (±4)	6 (±2)
$CH_2{=}CH_2$	+ 5.6	615 (±6)	6 (±2)
C_5H_5N	- 22.7	598 (±2)	6 (±2)
$\gamma\text{-}CH_3C_5H_4N$	- 23.0	594 (±2)	6 (±2)
$(C_6H_5)_3As$	0.0	548 (±2)	6 (±2)
CH_3CH_2NC	- 7.2	510 (±2)	
CH_3NC	- 7.1	510 (±2)	7 (±2)
CO	0.0	509 (±3)	6 (±2)
$:\overline{C^1C^2H_2C^3H_2C^4H_2O}$	- 10.1	395 (±2)	9 (±2)
$(CH_3)_2(C_6H_5)P$	+ 0.7	457 (±2)	

a) Data from Ref. [*241a*].

H References

1. Reed, F., Pregosin, P.S, Venanzi, L.M.: unpublished results
2. Crutchfield, M.M., Dungan, C.H., Letcher, J.H., Mark, V., Van Wazer, J.R., in: ^{31}P nuclear magnetic resonance. Topics in phosphorus chemistry, Vol 5. New York: Wiley-Interscience 1967
3. McFarlane, W., in: Determination of organic structures by physical methods, Vol 4, p 153. Nachod, F.C., Zuckerman, J.J.,(eds.). New York: Academic Press 1971
4. a) Farrar, T.C., Becker, E.D.: Pulse and Fourier transform NMR. New York: Academic Press 1971;
   b) Müllen, K., Pregosin, P.S.: Fourier transform NMR techniques: A practical approach. New York: Academic Press 1976;
   c) Shaw, D.: Fourier transform NMR spectroscopy, Elsevier, 1976
5. Ernst, R.R.: Sensitivity enhancement in magnetic resonance. In advances in magnetic resonance, Vol 2. New York: Academic Press, 1966
6. Schaeublin, S., Hoehener, A., Ernst, R.R.: J. Magn. Res. *13*, 196 (1974)
7. a) Koole, N.J., DeKoning, A.J., DeBie, N.J.A.: J. Magn. Res., *25*, 375 (1977);
   b) Gloneck, T., Wang, P.J., van Wazer, J.R.: J. Amer. Chem. Soc., *98*, 7968 (1976);
   c) Gloneck, T., van Wazer, J.C.: J. Phys. Chem., *80*, 639 (1976);
   d) Dale, S.W., Hobbs, M.E.: J. Phys. Chem., *75* 3537 (1971)
8. Lyerla, J.R., Levy, G.C., in: Topics in carbon-13 NMR spectroscopy (ed. Levy, G.C.) , Chap. 3. New York: Wiley Interscience, 1974
9. Pregosin, P.S.: unpublished observations. Although we have observed this effect in many phosphine complexes, the basic phenomena is well known. See: Bloembergen, N., Purcell, E.M., Pound, R.V.: Phys. Rev., *73*, 679 (1948)
10. Noggle, J.H., Schirmer, R.E.: The nuclear Overhauser effect, chemical applications. New York, London: Academic Press 1971
11. Solomon, J.: Phys. Rev., *99*, 559 (1955)
12. Kuhlman, K.F., Grant, D.: J. Amer. Chem. Soc., *90*, 7355 (1968)
13. Hawkes, G.E., Litchmann, W.M., Randall, E.W.: J. Magn. Res., *19*, 255 (1975)
14. Bösch, W., Pregosin, P.S.: submitted to Helv. Chem. Acta, 1979
15. a) Bierbeck, H., Martino, R., Saunders, J.K.: Canad. J. Chem., *54*, 1918 (1976);
    b) Yeagle, P.L., Hutton, W.C., Bruce Martin, R.: J. Amer. Chem. Soc., *97*, 7175 (1975)
16. Ramsey, N.F.: Phys. Rev., *91*, 303 (1953)
17. McConnell, H.M.: J. Chem. Phys., *24*, 460 (1956)

18. Pople, J.A., Santry, D.P.: Mol. Phys., *8*, 1 (1964)

19. Pople, J.A., Santry, D.P.: Mol. Phys., *9*, 311 (1965)

20. Gray, G.A., Cremer, S.E.: J. Chem. Soc., Chem. Comm., 451 (1974)

21. Jameson, C.J., Gutowsky, H.S.: J. Chem. Phys., *51*, 2790 (1969)

22. Cowley, A.H., White, W.D., Mannat, S.L.: J. Amer. Chem. Soc., *89*, 6433 (1967)

23. Rehder, D., Dorn, W.L.: Trans. Met. Chem., *1*, 233 (1976)

24. Rehder, D.: J. Magn. Res., *25*, 177 (1977)

25. Mann, B.E.: J.C.S., Chem. Comm., 1173 (1971)

26. Lucken, E.A.C., Noack, K., Williams, D.F.: J. Chem. Soc. (A) 148 (1967)

27. a) Keiter, R.L., Verkade, J.G.: Inorg. Chem., *8*, 2115 (1969);
b) Mc Farlane, W., Rycroft, D.S.: J. Magn. Reson., *24*, 95 (1976)

28. Milbrath, D.S., Verkade, J.G.: Inorg. Nucl. Chem. Letters, *12*, 921 (1976)

29. Mann, B.E., Masters, C., Shaw, B.L.: J. Chem. Soc. (A) 1104 (1971)

30. Grim, S.O., Satek, L.C.: J. Coord. Chem., *3*, 307 (1974)

31. Mann, B.E., Masters, C., Shaw, B.L.: J. Chem. Soc. Dalton 704 (1972)

32. Van Gaal, H.L.M., Verlaan, J.P.J.: J. Organomet. Chem., *133*, 93 (1977)

33. Muetterties, E.L., Alegranti, C.W.: J. Amer. Chem. Soc., *94*, 6386 (1972)

34. Mann, B.E.: Inorg. Nucl. Chem. Letters 7 595 (1971)

35. Goel, R.G., Ogini, W.O.: Inorg. Chem., *16*, 1968 (1977)

36. Kennedy, J.D., McFarlane, W., Wrackmeyer, B.: Inorg. Chem. *15*, 1299 (1976)

37. McFarlane, W., Rycroft, D.S.: J. Chem. Soc. Dalton, 1977 (1974)

38. Malone, G.F., Mann, B.E.: Inorg. Nucl. Chem. Letters *8*, 819 (1972)

39. McFarlane, W., Rycroft, D.S.: J. Chem. Soc., Chem. Comm., 336 (1973)

40. Grim, S.O., Wheatland, D.A. , McFarlane, W.: J. Amer. Chem. Soc., *89*, 5573 (1967)

41. Grim, S.O. et al.: J. Amer. Chem. Soc., *96*, 3416 (1974)

42. Brisdon, B.J.: J. Organomet. Chem., *125*, 225 (1977)

43. Fischer, E.O., Keiter, R.L., Krauss, L., Verkade, J.G.: J. Organomet. Chem., *37*, C7 (1972)

44. Gill, D.F., Mann, B.E., Masters, C., Shaw, B.L.: J. Chem. Soc., Chem. Comm. 1269 (1970)

45. Mather, G.G., Rapsey, G.J.N., Pidcock, A.: Inorg. Nucl. Chem. Letters *9*, 567 (1973)

46. Allen, F.H., Pidcock, A., Waterhouse, C.R.: J. Chem. Soc. (A) 2087 (1970)

47. Allen, F.H., Pidcock, A.: J. Chem. Soc. (A) 2700 (1968)

48. Socrates, G.: J. Inorg. Nucl. Chem. *31*1667 (1969)

49. Yamasaki, A., Fluck, E.: Z. anorg. allg. Chem., *396*, 297 (1973)

50. Grim, S.O., Pui Jun Lui, Keiter, R.L.: Inorg. Chem., *13*, 342 (1974)

51. Schmidbaur, H., Räthlein, K.H.: Chem. Ber., *106*, 2491 (1973)

52. a) see Appendix B
b) Nixon, J.F., Pidcock, A.: Annual review of NMR spectroscopy (ed. Mooney E.F.), Vol 2, p 357. London and N.Y., Academic Press, 1969

53. McFarlane, W., Rycroft, D.S.: J.C.S., Dalton 1977 (1974)

54. Mc Farlane, C., McFarlane, W., Rycroft, D.S.: J.C.S., Dalton 1616 (1976)

55. Goodman, B.A., Raynor, J.B.: J. Inorg. Nucl. Chem., *32*, 3406 (1970)

56. Nefedov, V.I., Yarzhemsky, V.G., Tarasov, V.P.: Chem. Phys. *18*, 417 (1976)

57. Allen, F.A., Sze, S.N.: J. Chem. Soc. (A) 2054 (1971)

58. Pidcock, A., Richards, R.E., Venanzi, L.M.: J. Chem. Soc. (A) 1707 (1966)

59. Schuhmann, H., Kroth, H.J.: Z. Naturforsch. *32b*, 768 (1977)

60. Vande Griend, L.J., Verkade, J.G., Pennings, J.Γ.M., Buck, H.M.: J. Amer. Chem. Soc., *99*, 2459 (1977)

61. Grim, S.O., Keiter, R.L., McFarlane, W.: Inorg. Chem., *6*, 1133 (1967)

62. Cowley, A.H., White, W.D.: J. Amer. Chem. Soc., *91*, 1913 (1969)

63. Sen, A., Halpern, J.: J. Amer. Chem. Soc., *99*, 8337 (1977)

64. Tolman, C.A.: Chem. Revs., *77*, 313 (1977)

65. Wayland, B.B., Abd-Elmageed, M.E., Mehne, L.F.: Inorg. Chem., *14*, 1456 (1975)

66. Wayland, B.B., Abd-Elmageed, M.E.: J. Amer. Chem. Soc. *96*, 4809 (1974)

67. Lehn, J.M., Munsch, B.: Mol. Phys., *23*, 91 (1972)

68. Hiller, I.H., Saunders, V.R.: Trans. Faraday. Soc., *66*, 2401 (1970)

69. Mather, G.G., Pidcock, A., Rapsey, G.J.N.: J. Chem. Soc., Dalton 2095 (1973)

70. Mason, R., Meek, D.W.: Angew. Chem., *90*. 195 (1978)

71. Mason, R., Randaccio, L.: J. Chem. Soc. (A) 1150 (1971)

72. Plastas, H.J., Stewart, J.M., Grim, S.O.: Inorg. Chem., *12*, 265 (1973)

73. Hitchcock, P.B., Jacobson, B., Pidcock, A.: J. Chem. Soc., Dalton 2043 (1977)

74. Hitchcock, P.B., Jacobson, B., Pidcock, A.: J. Chem. Soc., Dalton 2038 (1977)

75. Hitchcock, P.B., Jacobson, B., Pidcock, A.: J. Organomet. Chem., *136*, 397 (1977)

76. Brunett, M.J., Donaldson, P.B.: Inorg. Chem., *16*, 665 (1977)

77. Mather, G.C., Pidcock, A.: J. Chem. Soc. (A) 1226 (1970)

78. Appleton, T.G., Clark, H.C., Manzer, L.E.: Coordination Chem. Rev., *10*, 335 (1973)

79. Chatt, J., Mason, J.C.R., Meek, D.W.: J. Amer. Chem. Soc., *97*, 3826 (1975)

80. a) Grim, S.O., Del Gaudio, J., Molender, R.P., Tolman, C.A., Jesson, J.P.: J. Amer. Chem. Soc., *96*, 3416 (1974)
b) Grim, S.O., Mitchell, J.D.: Inorg. Chem., *16*, 1770 (1977)
c) Grim, S.O., Barth, R.C., Mitchell, J.D., Del Gaudio, J.: ibid, *16*, 776 (1977)
d) Grim, S.O., Briggs, W.L., Barth, R.C., Tolman, C.A., Jesson, J.P.: ibid, *13*, 1095 (1974)
e) Grim, S.O., Satek, L.C., Tolman, C.A., Jesson, J.P.: ibid, *14*, 656 (1975)
f) Grim, S.O., Mitchell, J.D.: ibid, *16*, 1762 (1977)

81. Appleton, T.G., Bennett, M.A., Tomkins, I.B.: J. Chem. Soc., Dalton 439 (1976)

82. Kunz, R.W., Pregosin, P.S., Bürgi, H.B.: to be published

83. Dingle, T.W., Dixon, K.R.: Inorg. Chem., *13*, 846 (1974)

84. Hammer, E.R., Kemmit, R.D.W., Smith, M.A.R.: J. Chem. Soc., Dalton 261 (1977)

85. Carty, A.J., Jakobson, S.E.: J. Chem. Soc., Chem. Comm., 175 (1975)

86. Carty, A.J.: Inorg. Chem., *15*, 1956 (1976)

87. Eaborn, C., Pidcock, A., Steele, B.R.: J. Chem. Soc., Dalton 767 (1976)

88. van Bliet, P.I., Kuyper, J., Vrieze, K.: J. Organomet. Chem., *122*, 99 (1976)

89. Kunz, R.W., Motschi, H., Pregosin, P.S.: unpublished results

90. Harris, R.K.: Can. J. Chem., *42*, 2275 (1964)

91. Bertrand, R.D., Ogilbie, F.B., Verkade, J.G.: J. Amer. Chem. Soc., *92*, 1908 (1970)

92. Verkade, J.G.: Coord. Chem. Rev., *9*, 1 (1972/73)

93. Lynden-Bell, R.M., Nixon, J.F., Schmutzler, R.: J. Chem. Soc. (A) 565 (1970)

94. Ogilbie, F.B., Jenkins, J.M., Verkade, J.G.: J. Amer. Chem. Soc., *92*, 1916 (1970)

95. deBie, M.J.A.: unpublished result 1973

96. Muetterties, E.L., Alegranti, C.W.: J. Amer. Chem. Soc., *94*, 6386 (1972)

97. Johnson, D.K., Pregosin, P.S., Venanzi, L.M: Helv. Chim. Acta, *59*, 2691 (1976)

98. Hollander, F.J., Ip, Y.L., Coucouvanis, D.: Inorg. Chem., *15*, 2230 (1976)

99. Grim, S.O., Schah, D.P.: Inorg. Nucl. Chem. Letters, *14*, 105 (1978)

100. Bennett, J., Pidcock, A., Waterhouse, C.R., Coggon, P., McPhail, A.T.: J. Chem. Soc. (A) 2094 (1970)

101. Mann, B.E.: Inorg. Nucl. Chem. Letters, *7*, 595 (1971)

102. Goel, R.G., Ogini, W.O.: Inorg. Chem. *16*, 1968 (1977)

103. Meakin, P., Guggenberger, L.J., Tebbe, F.N., Jesson, J.P.: Inorg. Chem., *13*, 1025 (1974)

104. Keiter, R.L., Verkade, J.G.: Inorg. Chem., *8*, 2115 (1969)

105. Grim, S.O., McAllister, P.R., Singer, R.M.: J. Chem. Soc., Chem. Comm., 38 (1969)

106. Grim, S.O., Wheatland, D.A.: Inorg. Chem., *8*, 1716 (1969)

107. Mathieu, R., Lenzi, N., Poilblanc, R.: Inorg. Chem., *9*, 2030 (1970)

108. a) George, T.A., Turnipseed, C.D.: Inorg. Chem., *12*, 394 (1973);
b) George, T.A., Sterner, C.D.: ibid, *15*, 165 (1976)

109. Taylor, R.C., Keiter, R.L., Cary, L.W.: Inorg. Chem., *13*, 1928 (1974)

110. Keiter, R.L., Fasig, K.M., Cary, L.W.: Inorg. Chem., *14*, 201 (1975)

111. Dawson, J.W., Venanzi, L.M.: J. Amer. Chem. Soc., *90*, 7229 (1968)

112. Mynott, R.J., Pregosin, P.S., Venanzi, L.M.: J. Coord. Chem., *3*, 145 (1973)

113. Hyde, E.M., Kennedy, J.D., Shaw, B.L., McFarlane, W.: J.C.S. Dalton 1571 (1977)

114. Grim, S.O., Ference, R.A.: Inorg. Chim. Acta *4*, 277 (1970)

115. Mann, B.E., Masters, C., Shaw, B.L.: J.C.S. Dalton 704 (1972)

116. a) Brown, T.H., Green, P.J.: J. Amer. Chem. Soc., *92*, 2359 (1970)
b) Tolman, C.A., Meakin, P.R., Lindner, D.L., Jesson, P.J.: J. Amer. Chem. Soc., *96*, 2762 (1974)

117. a) Garrou, P.E., Hartwell, G.E.: Inorg. Chem., *15*, 646 (1976);
b) Van Vliet, P.I., Vrienze, K.: J. Organomet. Chem., *139* 337 (1977)
c) Sanger, A.R.: J.C.S. Dalton 120 (1977)

118. Bright, A., Mann, B.E., Masters, C., Shaw, B.L., Stade, R.M., Stainbank, E.: J. Chem. Soc. A 826 (1971); Mann, B.E., Masters, C., Shaw, B.L., Stainbank, R.E.: Chem. Commun. 1103 (1971)

119. Empsall, H.D., Hyde, E.M., Jones, C.E., Shaw, B.L.: J. Chem. Soc., Dalton 1980 (1974)

120. Meakin, P., Jesson, J.P.: J. Amer. Chem. Soc., *95*, 7272 (1973)

121. Garrou, P.E., Curtis, J.L.S., Hartwell, G.E.: Inorg. Chem., *15*, 3094 (1976)

122. Lau, K.S., Becker, Y., Huang, F., Baenzinger, N., Stille, J.K.: J. Amer. Chem. Soc., *99*, 5664 (1977)

123. Egglestone, D., Baird, M.C., Lock, C.J.L., Turner, G.: J.C.S. Dalton 1576 (1977)

124. Tiethof, J., Peterson, J.L., Meek, D.W.: Inorg. Chem., *15*, 1356 (1976)

125. Gaulton, K.C.: Inorg. Chem., *13*, 1774 (1974)

126. Eaborn, C., Farrell, N., Murphy, J.L., Pidcock, A.: J.C.S. Dalton 58 (1976)

127. Van Gaal, H.L.M., Verlaak, J.M.J., Posno, T.: Inorg. Chim. Acta. *23*, 43 (1977)

128. Huis, R., Masters, C.: J.C.S. Dalton 1796 (1976)

129. Venanzi, L.M.: Chemistry in Britain, *4*, 162 (1968)

130. Hammer, E.R., Kemmit, R.D.W., Smith, M.A.R.: J. Chem. Soc., Dalton 261 (1977)

131. Appleton, T.G., Bennett, M.A., Tomkins, I.B.: J. Chem. Soc., Dalton 439 (1976)

132. Slack, D.A., Baird, M.C.: Inorg. Chim. Acta, *24*, 277 (1977)

133. Ahmad, N., Ainscough, E.W., James, T.A., Robinson, S.D.: J.C.S. Dalton 1151 (1973)

134. Grim, S.O., Keiter, R.L., McFarlane, W.: Inorg. Chem., *6*, 1133 (1967)

135. Kennedy, J.D., McFarlane, W., Puddephatt, R.J.: J.C.S. Dalton 745 (1976)

136. Kennedy, J.D., McFarlane, W., Puddephatt, R.J., Thompson, P.J.: J.C.S. Dalton 874 (1976)

137. Dixon, K.R., Fakley, M., Pidcock, A.: Can. J. Chem., *54*, 2733 (1976)

138. Allen, F.H., Pidcock, A.: J. Chem. Soc. A, 2700 (1968)

139. Kreutzer, P.H., Tschörpp, K.T., Beck, W.: Z. Naturforsch. *306*, 544 (1975)

140. Dilton, K.B., Waddington, T.C., Younger, D.: J. Chem. Soc., Dalton 790 (1975)

141. Brown, M.P., Puddephatt, R.J., Rashidi, M., Seddon, K.R.: J. Chem. Soc., Dalton 951 (1977)

142. a) Ahmad, N., Aiscough, E.W., James, T.A., Robinson, S.D.: J.C.S. Dalton 1148 (1973)
b) Meakin, P., Jesson, J.P.: J. Amer. Chem. Soc., *96*, 5751 (1974)

143. Pregosin, P.S., Sze, S.N.: Helv. Chim. Acta, *60*, 1371 (1977)

144. Grosse, J., Schmutzler, R.: J.C.S. Dalton 405 (1970)

145. Crocker, C., Goodfellow, R.J.: J.C.S. Dalton 1687 (1977)

146. Duddell, D.A., Evans, J.G., Goggin, P.L., Goodfellow, R.J., Rest, A.J., Smith, J.G.: J. Chem. Soc. A 2134 (1969)

147. Goggin, P.L., Goodfellow, R.J., Knight, J.R., Norton, M.G., Taylor, B.F.: J.C.S. Dalton 2220 (1973)

148. Goodfellow, R.J., Taylor, B.F.: J.C.S. Dalton 1676 (1974)

149. Sze, S.N., Pregosin, P.S.: submitted for publication in Helv. Chim. Acta (1978)

150. Cheney, A.J., Mann, B.E., Shaw, B.L., Slade, R.N.: J. Chem. Soc. A 3833 (1971); Jones, C.E., Shaw, B.L., Turtle, B.L.: J.C.S. Dalton 992 (1974)

151. Steele, B.R., Vrieze, K.: Trans. Met. Chem. *2*, 169 (1977);
Ros, R., Michelin, R.A., Bataillard, R., Roulet, R.: J. Organomet. Chem. *139*, 355 (1977);

Fakley, M., Pidcock, A.: J.C.S. Dalton 1444 (1977)

152. Cardin, C.J., Cardin, D.J., Lappert, M.: J.C.S. Dalton 767 (1977)

153. Garrou, P.E., Heck, R.F.: J. Amer. Chem. Soc. *98*, 4115 (1976)

154. Miyamoto, J.: J. Organomet. Chem., *134*, 335 (1977)
Socrates, G.: J. Inorg. Nucl. Chem., *31*, 1667 (1969)

155. Carty, A.J.: Inorg. Chem., *15*, 1956 (1976)

156. Sanger, A.R.: J.C.S. Dalton 1971 (1977)

157. Empsall, H.D., Hyde, E.M., Mentzer, E., Shaw, B.L.: J.C.S. Dalton 2285 (1977)

158. Goel, R.G., Montemayor, R.G.: Inorg. Chem. *16*, 2183 (1977)

159. McFarlane, W.: Proc. Royal Soc. A, *306*, 185 (1968)

160. McFarlane, W., Rycroft, D.S.: J.C.S. Dalton 1977 (1974)

161. Schumann, H., Kroth, H.J.: Z. Naturforsch., *32b*, 876 (1977)

162. Crosbie, K.B., Sheldrick, D.M.: Mol. Phys. *20*, 317 (1971);
Fritz, G., Schöfer, H.: Z. anorg. allg. Chem., *409*, 137 (1974)

163. Verkade, J.G., King, R.W., Heitsch, C.W.: Inorg. Chem., *3*, 884 (1964)

164. Schmutzler, R.: Adv. Fluorine Chem., *5*, 255 (1965)

165. Rudolf, R.W., Parry, R.W.: J. Amer. Chem. Soc. *89*, 1621 (1967)

166. Cowley, A.W., Damasco, M.C.: J. Amer. Chem. Soc., *93*, 6815 (1971)

167. Rudolph, R.W., Schultz, C.W.: J. Amer. Chem. Soc., *93*, 6821 (1971)

168. Vrieze, W.H.N., Jellinek, F.: Chem. Phys. Lett., *1*, 284 (1967)

169. Gray, G.A., Albright, T.A.: J. Amer. Chem. Soc., *99*, 3243 (1977)

170. Stee, W.J., Okruszek, A., Micholski, J.: Phosphorus, *2*, 97 (1972)

171. Verkade, J.G.: Coord. Chem. Revs., *9*, 1 (1972/73)

172. Goodfellow, R.J., Taylor, B.F.: J. Chem. Soc., Dalton 1676 (1974)

173. Fornies, J., Green, M., Spencer, J.L., Stone, F.G.: J.C.S. Dalton 1006 (1977)

174. Gerlach, D.H., Kane, A.R., Parshall, G.W., Jesson, J.P., Muetterties, E.L.: J. Amer. Chem. Soc., *93*, 3543 (1971)

175. Grove, D., Venanzi, L.M.: unpublished results, ETH-Zürich 1977

176. a) Bracher, G., Venanzi, L.M.: unpublished results, ETH-Zürich 1977
b) Boron, P., Venanzi, L.M.: unpublished results, ETH-Zürich 1977

177. Tolman, C.A., Seidel, W.C., Gerlach, D.H.: J. Amer. Chem. Soc., *94*, 2669 (1972)

178. Mann, B.E., Shaw, B.L., Tucker, N.I.: J. Chem. Soc. (A) 2667 (1971)

179. Cairns, M.A., Dixon, K.R., McFarlane, J.J.: J.C.S. Dalton 1159 (1975)

180. Crocker, C., Goodfellow, R.J., Gimens, J., Uson, R.: J.C.S. Dalton 1448 (1977)

181. Reed, F.J.S., Venanzi, L.M.: Helv. Chim. Acta, *60*, 2804 (1977)

182. Mann, B.E., Masters, C., Shaw, B.L.: J.C.S. Dalton 48 (1972)

183. Karplus, M., Pople, J.A.: J. Chem. Phys. *38*, 2803 (1963)
Saika, A., Slichter, C.P.: J. Chem. Phys., *22*, 26 (1954)
Ramsey, N.F.: Phys. Rev., *86*, 243 (1952)

184. a) Letcher, J.H., Van Wazer, J.A.: J. Chem. Phys., *45*, 2916 (1966);
b) ibid, *44*, 815 (1966);
c) ibid, *45*, 2926 (1966)

185. Purdella, D.: J. Magn. Res., *5*, 23 (1971); Grim, S.O., Plastes, H.J., Huheey, C.L., Huheey, J.E.: Phosphorus, *1*, 61 (1971)

186. Gorenstein, D.G.: J. Amer. Chem. Soc., *97*, 898 (1975)

187. a) Jongsma, C., DeKleijn, J.P., Bickelhaupt, F.: *Tetrahedron*, *30*, 3465 (1974)
b) Darensburg, M.Y., Daigle, D.: Inorg. Chem., *14*, 1217 (1975)

188. Schumann, H., Kroth, H.J.: Z. Naturforsch., *32b*, 513 (1977)

189. Grim, S.O., McFarlane, W., Davidoff, E.F., Marks, T.J.: J. Phys. Chem. *70*, 581 (1966); Grim, S.O., McFarlane, W., Davidoff, E.F.: J. Org. Chem., *32*, 781 (1967); Grim, S.O., McFarlane, W.: Can. J. Chem., *46*, 2071 (1968)

190. Quinn, L.D., Breen, J.J.: Org. Magn. Res., *5*, 17 (1973)

191. Maciel, G., in: Topics in carbon-13 NMR spectroscopy (ed. Levy, G.C.), Vol 1 Chap. 2. New York: Wiley-Interscience 1974

192. Pregosin, P.S., Randall, E.W., White, A.I.: Chem. Comm. 1602 (1971)

193. Balimann, G., Pregosin, P.S.: J. Magn. Res. *26*, 283 (1977)

194. Mann, B.E., Masters, C., Shaw, B.L., Slade, R.M., Stainbank, R.E.: Inorg. Nucl. Chem. Letters, *7*, 881 (1971)

195. Kunz, R.W., Pregosin, P.S.: unpublished results

196. Balimann, G., Pregosin, P.S.: J. Magn. Res., *22*, 235 (1976)

197. Chaudret, B.N., Cole-Hamilton, D.J., Nohr, R.S., Wilkinson, G.: J.C.S. Dalton 1546 (1977)

198. Verstuyft, A.W., Nelson, J.H., Cary, L.W.: Inorg. Nucl. Chem. Letters, *12*, 53 (1976); Verstuyft, A.W., Cary, L.W., Nelson, J.H.: Inorg. Chem., *14*, 1495 (1975)

199. Mann, B.E., Musco, A.: J.C.S. Dalton 1673 (1975)

200. Grim, S.O., Keiter, R.L.: Inorg. Chim. Acta., *4*, 56 (1970)

201. Hoffman, P.R., Gaulton, K.G.: J. Amer. Chem. Soc., *97*, 4221 (1975)

202. Dobson, A., Robinson, S.D.: Inorg. Chem., *16*, 1321 (1977)

203. Garrou, P.E.: Inorg. Chem., *14*, 1435 (1975)

204. King, R.B., Cloyd Jr. J.C.: Inorg. Chem., *14*, 1550 (1975)

205. Pregosin, P.S., Kunz, R.W., Kapoor, P.N., Venanzi, L.M.: Manuscript in preparation.

206. Gill, D.F., Mann, B.E., Shaw, B.L.: J.C.S. Dalton 311 (1973)

207. Mathieu, R., Lenzì, M., Poilblanc, R.: Inorg. Chem., *9*, 2030 (1970)

208. Tolman, C.A.: J. Amer. Chem. Soc., *92*, 2956 (1970)

209. Karsch, H.H.: Angew. Chemie, *89*, 57 (1977)

210. Sze, S.N., Pregosin, P.S.: accepted for publication Helv. Chim. Acta. 1978

211. Schmidtbaur, H., Gasser, O., Fraser, T.E., Ebsworth, E.A.V.: J.C.S. Chem. Comm. 334 (1977)

212. van Vliet, P.I., Vrieze, K.: J. Organomet. Chem., *139*, 337 (1977)

213. Brown, M.P., Puddephatt, R.J., Rashidi, M., Seddon, K.R.: J.C.S. Dalton 516 (1978)

214. Vriends, R.C.J., van Koten, G., Vrieze, K.: Inorg. Chim. Acta, *26*, L29 (1978)

215. Hietkamp, S., Stufkens, D.J., Vrieze, K.: J. Organomet. Chem., *122*, 419 (1976)

216. Stewart Jr., R.P., Isbrandt, L.R., Benedict, J.J.: Inorg. Chem., *15*, 2013 (1976)

217. Venanzi, L.M.: Chimia, *30*, 227 (1976)

218. DuBois, D.L., Meek, D.W.: Inorg. Chem., *15*, 3076 (1976)

219. Jesson, J.P., Muetterties, E.L., in: Dynamic nuclear magnetic resonance spectroscopy. (ed. Jackman, L.M., Cotton, F.A.) p 253. New York: Academic Press Inc. 1975

220. a) English, A.D., Ittel, S.D., Tolman, C.A., Meakin, P., Jesson, J.P.: J. Amer. Chem. Soc., *99*, 117 (1977)
b) Meakin, P., Jesson, J.P.: ibid, *96*, 5760 (1974)

221. Pregosin, P.S., Venanzi, L.M.: Helv. Chim. Acta, *58*, 1548 (1975)

222. Mann, B.E.: J.C.S. Perkin II 30 (1972)

223. McFarlane, W.: Proc. Royal Soc. A, *306*, 185 (1968)

224. Gray, G.A., Cremer, S.E.: J.C.S. Chem. Comm., 451 (1974)

225. Jenkins, J.M., Shaw, B.L.: J. Chem. Soc., A 770 (1966)

226. Axelson, D.E., Holloway, C.E.: J.C.S. Chem. Comm. 455 (1973)

227. Aktar, M., Ellis, P.D., Mac Diarmid, A.G., Odom, J.D.: Inorg. Chem., *11*, 2917 (1972)

228. Redfield, D.A., Cary, L.W., Nelson, J.H.: Inorg. Chem., *14*, 50 (1975); ibid, *15*, 732 (1976); ibid, *15*, 1128 (1976)

229. a) Pregosin, P.S., Kunz, R.W.: Helv. Chim. Acta, *58*, 423 (1975);
b) Balimann, G., Motschi, H., Pregosin, P.S.: Inorg. Chim. Acta, *23*, 191 (1977)

230. Kuran, W., Musco, A.: Inorg. Chim. Acta, *12*, 187 (1975)

231. Verstuyft, A.W., Nelson, J.H., Cary, L.W.: Inorg. Chem., *15*, 732 (1976)

232. a) Van Baar, J.F., Vrieze, K., Stufkens, D.J.: J. Organomet. Chem. *85*, 249 (1975)
b) Stewart Jr. R.P., Isbrandt, L.R., Benedict, J.J., Palmer, J.G.: J. Amer. Chem. Soc., *98*, 3215 (1976)

233. Dias, S.A., Downs, A.W., McWhinnie, W.R.: J.C.S. Dalton 162 (1975)

234. Dehand, J., Pfeffer, M.: Coord. Chem. Rev., *18*, 327 (1976)

235. Derek, H., Clague, H., Masters, C.: J.C.S. Dalton 858 (1975)

236. Tolman, C.A., English, A.D., Manzer, L.E.: Inorg. Chem., *14*, 2353 (1975)

237. Bodner, G.M.: Inorg. Chem., *14*, 1932 (1975)

238. Coulson, D.R.: J. Amer. Chem. Soc., *98*, 3111 (1976)

239. Cherwinski, W.J., Johnson, B.F.G., Lewis, J., Norton, J.R.: J.C.S. Dalton 1156 (1975)

240. Anglin, J.R., Calhoun, H.P., Graham, W.A.G.: Inorg. Chem., *16*, 2281 (1977)

241. a) Chisholm, M.H., Clark, H.C., Manzer, L.E., Stothers, J.B., Ward, J.E.H.: J. Amer. Chem. Soc., *95*, 8574 (1973);
b) Clark, H.C., Ward, J.E.H.: Ibid, *96*, 1741 (1974);
c) Clark, H.C., Ward, J.E.H.: Can. J. Chem., *52*, 570 (1974);
d) Clark, H.C., Manzer, L.E., Ward, J.E.H.: Ibid, *52*, 1165 (1974);
e) Clark, H.C., Manzer, L.E., Ward, J.E.H.: Ibid, *52*, 1973 (1974);
f) Chisholm, M.H., Clark, H.C., Manzer, L.E., Stothers, J.B., Ward, J.E.H.: J. Amer. Chem. Soc., *97*, 721 (1975);
g) Chisholm, M.H., Clark, H.C., Ward, J.E.H., Yasufuku, K.: Inorg. Chem., *14*, 894 (1975);
h) Clark, H.C., Ward, J.E.H., Yasufuku, K.: Can. J. Chem. *53*, 186 (1975)

242. Mann, B.E., in: Advances in organometallic chemistry. Vol 12, p 135; New York: Academic Press, 1974; Chisholm, M.H., Godleski, S., in: Progress in inorganic chemistry, (ed. Lippard, S.J.) Vol 20, p 299. New York: John Wiley 1976; Gansow, O.A., Vernon, W.D., in: Topics in carbon-13 NMR spectroscopy (ed. Levy, G.C.), Vol 2, p 270. New York: Wiley Interscience.

243. Desclaux, J.P.: Atomic data and nuclear data tables, *12*, 311 (1973)

244. Basch, H., Gray, H.B.: Inorg. Chem., *6*, 365 (1967)

245. Kroening, R.F., Rush, R.M., Martin Jr., D.S., Clardy, J.C.: Inorg. Chem., *13*, 1366 (1974)

246. Interante, L.V., Messmer, R.P.: Chem. Phys. Letters, *26*, 225 (1974)

247. Kroening, R.F., Hunter, L.D., Rush, R.M., Clardy, J.C., Martin Jr., D.S.: J. phys. Chem., *77*, 3077 (1973)

248. Ho, H., Fujitu, J., Saito, K.: Bull. Chem. Soc. Japan, *40*, 2584 (1967)

249. Jørgensen, C.K.: Acta Chem. Scand., *10*, 518 (1956)

250. Jørgensen, C.K.: Orbitals in atoms and molecules. New York, London: Academic Press 1962

251. Jørgensen, C.K.: Absorbtions spectra and chemical bonding in complexes. Pergamon Press 1962

252. Pidcock, A., Richards, R.E., Venanzi, L.M.: J. Chem. Soc. (A), 1970 (1968)

253. Jørgensen, C.K.: Modern aspects of ligand field theory. North Holland Publishing Company, Amsterdam, London, 1971

254. Goodman, B.A., Raynor, J.B.: J. inorg. nucl. Chem., *32*, 3406 (1970)

255. Mann, J.B., Waber, J.T.: Atomic Data, *5*, 201 (1973)

256. Froese-Fischer, C.: Atomic Data, *4*, 301 (1972)

257. Mann, J.B.: Atomic data and nuclear data tables, *12*, 1 (1973)

258. Clementi, E., Roetti, C.: Atomic data and nuclear data tables, *14*, 177 (1974)

259. Basch, H., Gray, H.B.: Theoret. Chim. Acta (Berl.), *4*, 367 (1966)

Index

MIX
Papier aus verantwortungsvollen Quellen
Paper from responsible sources
FSC® C105338

If you have any concerns about our products,
you can contact us on
ProductSafety@springernature.com

In case Publisher is established outside the EU,
the EU authorized representative is:
Springer Nature Customer Service Center GmbH
Europaplatz 3, 69115 Heidelberg, Germany

Printed by Libri Plureos GmbH
in Hamburg, Germany